COAL MINE DISASTERS
— OF —
NORTH CAROLINA

COAL MINE DISASTERS
— OF —
NORTH CAROLINA

JOHN HAIRR

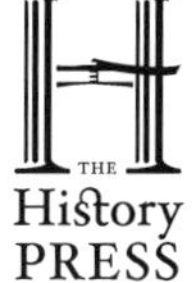

Published by The History Press
Charleston, SC
www.historypress.net

Cover images courtesy of the Ben Dixon McNeill Photographic Collection, North Carolina Collection Photographic Archives, the Wilson Library, University of North Carolina–Chapel Hill.

First published 2017

Manufactured in the United States

ISBN 978.1.46713.581.8

Library of Congress Control Number: 2016950700

CONTENTS

ACKNOWLEDGEMENTS

Much work researching this book was put in at the places one normally consults when writing a work of history, such as archives and libraries across the country, and I acknowledge the help given by many people at these institutions. From the smallest fragment of newspaper to the largest obscure tome of mining, the information gained has been very important in helping illuminate the history of the coal mining industry in North Carolina.

One of the main sources of information in a work of this nature is gained in the field, visiting the sites of old mines and related operations. By actually studying such sites on the ground, one is able to piece together a better understanding of the difficulties confronting those who in years gone by attempted to extract the minerals from the earth. This is especially true on the Cape Fear and Deep Rivers, where obstacles and impediments presented some unique challenges to getting the commodities to market.

Two individuals who are unfortunately no longer with us helped not so much with the writing of this particular book but instead with earlier field research into the history and geography of the Deep River region, and I would be remiss if I did not acknowledge their help. The first was historian and geologist Wade Hadley, who took the time to share with me much of his ideas and findings relating to the mining operations at Gulf and Coal Glen. Fortunately for historians and researchers, Mr. Hadley published a few articles on the topic, as well as a book covering the history of the Cape Fear & Deep River Navigation Company. Although his health prevented him from tramping through the woods to some of the remote sites in his

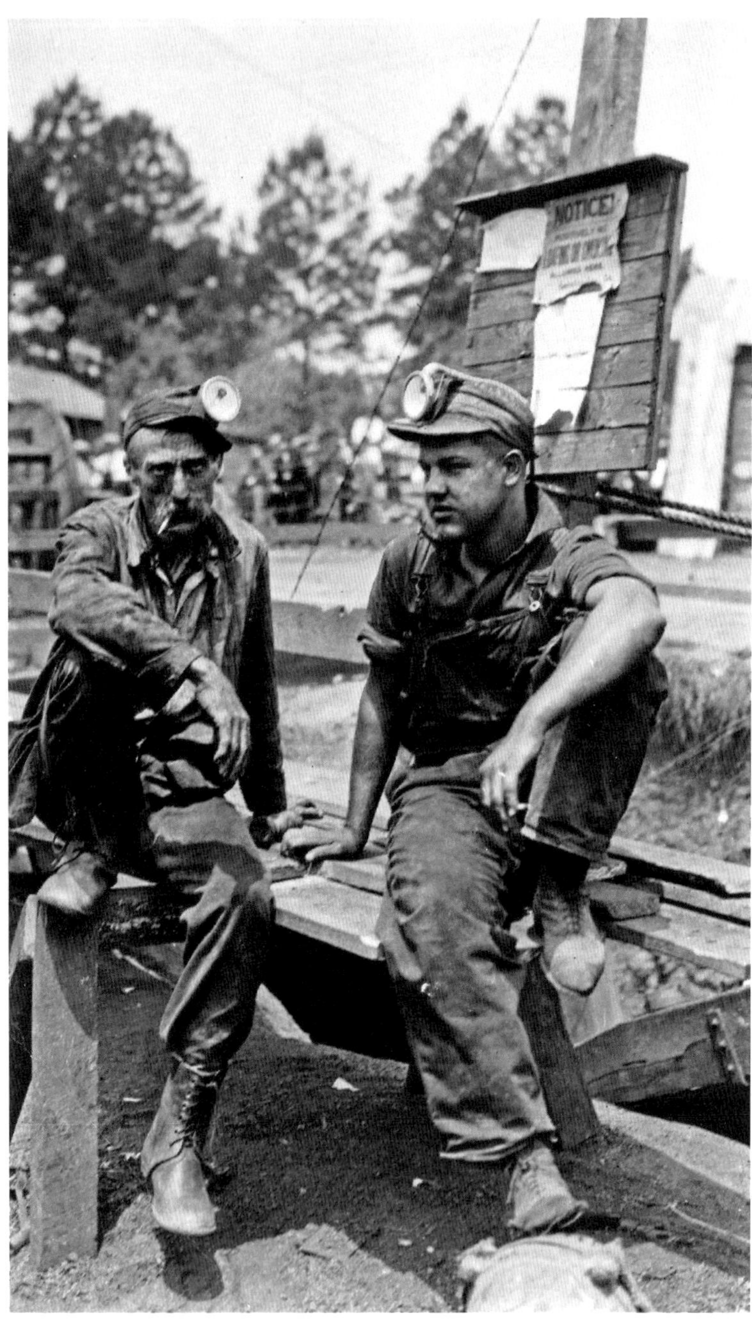

Miners taking a break from their rescue efforts at Coal Glen in Chatham County. *Ben Dixon McNeill Photographic Collection, North Carolina Collection Photographic Archives, Wilson Library, University of North Carolina–Chapel Hill.*

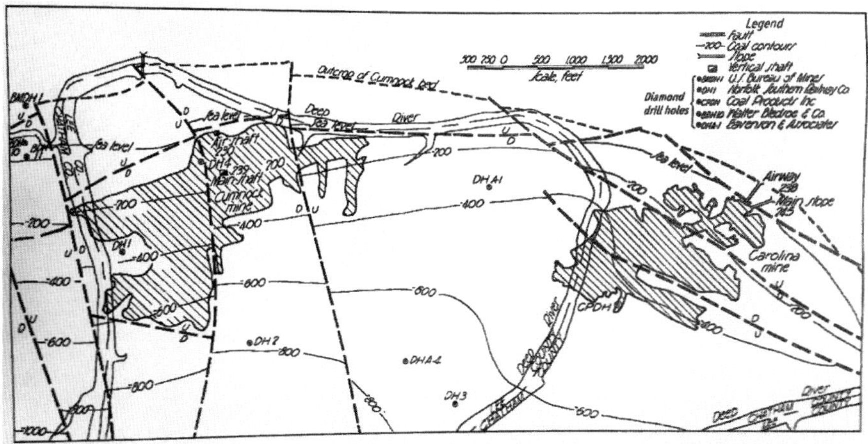

Map prepared by the U.S. Bureau of Mines in 1952 shows the proximity of the Cumnock and Coal Glen mining operations. *Author's collection.*

later years, he was especially fond of showing off the remains of the mining operations at Ore Hill in his native Chatham County.

The second individual who is no longer with us was my friend and colleague Guy Smith, who, among other things, served as the site manager at House in the Horseshoe State Historic Site near Carbonton. Guy and I spent many hours rambling across the hills of the Deep River country looking for traces of old coal and iron operations. Many times guided by an old map that one of us had dug up in the archives or some such collection, and armed with machete and compass, we ventured forth into some of the most inaccessible places along the Cape Fear and its tributaries, looking for piles of rocks, ore or slag that remain from the days when people hoped to make a fortune extracting and refining what lies hidden beneath the surface. We were able to publish an article in the *Cape Fear Journal* about the coal and iron operations in central North Carolina. We often discussed coauthoring a book on North Carolina's coal and iron industry, but Guy became ill and passed before that task was accomplished.

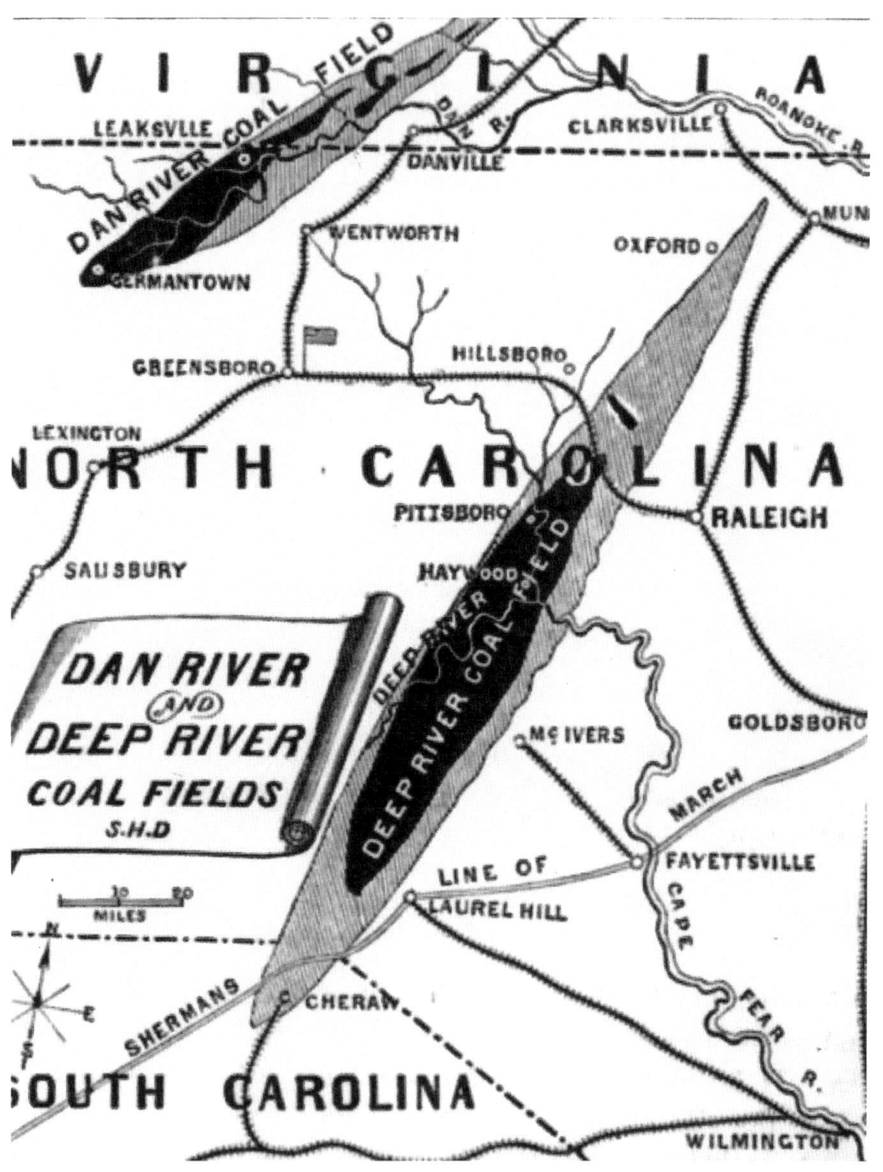

An 1866 map of the coal deposits found in North Carolina. *Author's collection.*

Chapter 1

ORIGINS OF NORTH CAROLINA'S COAL INDUSTRY

North Carolina has a long history of coal mining, dating back more than two centuries. John Willcox opened pit mines along the north side of the Deep River to fuel his iron operations at Gulf as early as 1768. These were surface pits where miners worked with picks, shovels and wheelbarrows on surface outcroppings—not the deep, subterranean operations that usually come to mind when thinking of coal mines. From these modest beginnings, North Carolina's coal miners eventually employed state-of-the-art technology to help raise more than 1 million tons of coal from deep below the surface of the earth.

There were several coal mines scattered along both sides of the Deep River in Moore, Lee and Chatham Counties. Many of these mines have long since been forgotten, including the Gulf Mine, the Deep River Mine, the McIver Mine, the Black Diamond Mine, the Murchison Mine and the Gardner Mine. Despite its reputation for being a dangerous place to mine coal, the Deep River Coal Field was not as deadly a place to work as some might suspect. In fact, for more than a century, miners descended deep below the surface of the earth and worked without major disasters at all the mines of the Deep River Coal Field except two—Egypt and Farmville.

An investigation of the history of the mining disasters does not satisfactorily answer the question of why these particular mines were so deadly. All of the mines in the Deep River Coal Field penetrated basically the same strata of the Triassic Age rock formations. The coal was extracted from two beds: the Gulf Coal Bed and the Cumnock Coal Bed, with the vast majority of coal being extracted from the latter. Both of these coal beds are located north and west of the city of Sanford.

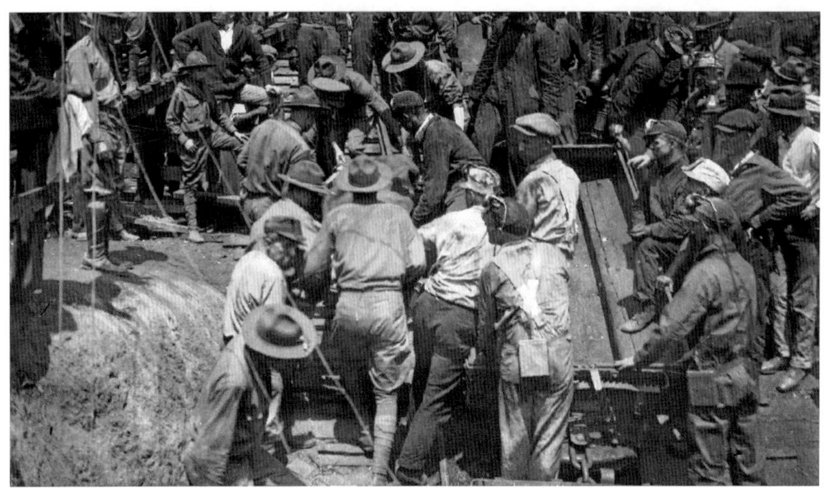

Rescue work at the Carolina Mine in May 1925. News and Observer *photo, courtesy North Carolina Archives and History.*

Looking at these disasters provides insights into how the mines were operated and helps to disprove many of the rumors that the mines were operated by profiteers and shysters more concerned with swindling a quick buck from the folks who lived in central North Carolina in years past than in producing coal. This explanation might have been valid had the same folks owned and operated the mines for the hundred or so years they were active, but as will be shown later in this work, this was not the case. In fact, some pretty intelligent individuals were involved in these mining operations, many with a strong background in mining safety and techniques learned from operations in Alabama and Pennsylvania. There were even mine operators from Cornwall and Scotland who brought much mining expertise from Europe to the Deep River region.

Coal mining is a dangerous occupation and has been for centuries. We will never know just how many people have perished in mining operations down through the years all over the world in mines from causes such as cave-ins, floods, runaway mining cars and explosions of gas. Official statistics kept by the Mine Safety and Health Administration of the U.S. Department of Labor point out that between 1900 and 2014, in the United States alone, 104,832 people were officially recorded as being killed in the coal mining business, with more than 90,000 of those deaths occurring between the years 1900 and 1945.

Clarence Hall and Walter O. Snelling noted that during the last decade of the nineteenth century, 9,974 people were killed in the coal mines across the United States. This was a particularly deadly decade for mining-related fatalities and was believed to be the deadliest in history.[1]

The high rate of fatalities continued into the first decade of the twentieth century. In 1907, during a month known as "Bloody December," 703 miners were killed in five separate mine explosions across the United States, the deadliest being the explosion at the Monogah Mine in West Virginia that killed 363 miners.[2]

This particularly brutal period led to Congress creating the U.S. Bureau of Mines in 1910. The man chosen by President William Taft to run the new organization was Joseph A. Holmes, who, among other posts, served as head of the North Carolina Geological Survey from 1891 to 1905. This was a particularly deadly period in North Carolina's mining history, with numerous tragic episodes occurring. Holmes drew on his experiences in the Old North State as he took over responsibility for devising a program of mining safety for the entire country. He worked passionately in his effort to improve the safety of mining operations and was a pioneer in many early experiments to try to ascertain the causes of explosions in coal mines. To this day, Holmes is still referred to as the "Father of the Bureau of Mines."[3]

Thanks to a scarcity of records, especially in the early years of the mining operations, there is no way of knowing how many workers have perished in North Carolina coal mines. Details of several accidents, as well as the names of many of those who died, in the Deep River Coal Field mining coal during the War Between the States have not survived. Nor do we know the tally of the poor souls who died in less than spectacular accidents in the mines in the days before official reports had to be filed—or before a newspaperman happened to be around to chronicle the event.

There are even discrepancies in the official numbers when they were reported. For instance, the official death toll for the 1925 Coal Glen explosion was fifty-three, a number transcribed shortly after the event. But as time passed, the bodies of more unfortunate miners were discovered, and by the fall of 1925, there were seventy-one bodies, with more still being reported as being entombed in the mine. The reason for this discrepancy is because mine officials were unsure of exactly how many people were at work in the mine at the time of the explosion.

Today, there is renewed interest in exploiting the minerals of the Deep River Coal Field. Geologists with laptops and lasers search places where, in years gone by, their predecessors probed with picks and shovels, each eager to find out what lies deep beneath the surface. For this reason, it seems appropriate that we study these early mining activities in order to learn all we can about the mines and the people who were affected by past attempts to extract these resources from the earth.

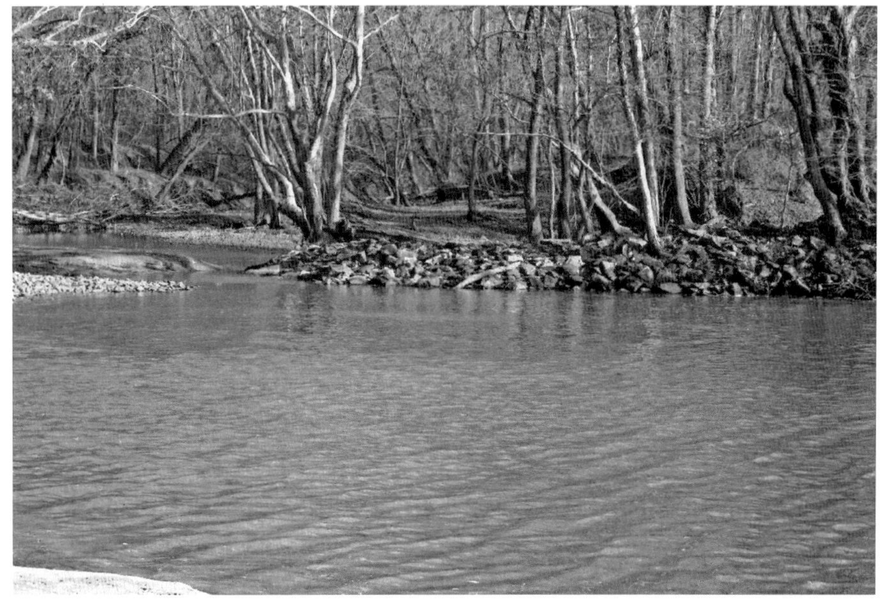

Keeping the Cape Fear and Deep Rivers open for navigation by steamboats was vital to the success of the plans of the development of the mineral resources of the Deep River Coal Field. Shown here are the remains of one of the locks and dams constructed by the Cape Fear & Deep River Navigation Company at Northington's Falls near Raven Rock in Harnett County. *Photo by the author.*

Chapter 2

ANTEBELLUM COAL ACTIVITIES AND EARLY MINING DISASTERS

The property where the Egypt Coal Mine was sunk was located in Chatham County on land that was part of Peter Evans's plantation. The plantation was originally known as Lagrange, but Evans changed the name to Egypt after a local farmer who was headed toward Evans's gristmill contrasted his lush river bottomlands with nearby drought-stricken farms in that area and likened Evans's lands with Egypt as described in the Old Testament story of Joseph.

Evans made some efforts to work the surface outcroppings of coal he found on his lands but left the area in the 1830s before a shaft was sunk underground. Evans sold the Egypt plantation in 1851 to Brooks Harris and Lawrence J. Haughton, with Harris subsequently buying out his partner. Harris set about the systematic examination of the coal and other mineral resources that he was certain lay just below the surface of his property on the south side of the Deep River in what is today Lee County; he also commenced work on sinking the exploratory shaft that became the Egypt Coal Mine. Geologists Marius Campbell and Kent Kimball later described Harris's efforts as "probably the most important single piece of development work ever undertaken in this coal field."[4]

Perhaps of equal importance to his decision to sink the shaft at Egypt was Harris's decision to bring in William McClane to oversee the operations. A mining expert who hailed from Pennsylvania, McClane was one of the most influential individuals in the development of the mineral resources of the Deep River country in the nineteenth century. In addition to his work

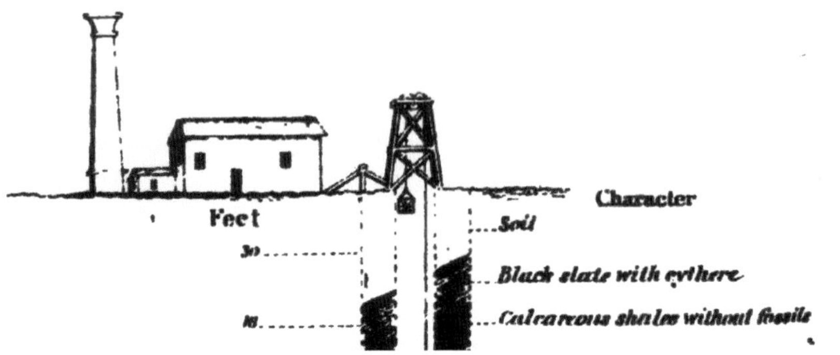

Close-up view of the structures on the surface of the Egypt Coal Mine in 1858. *Author's collection.*

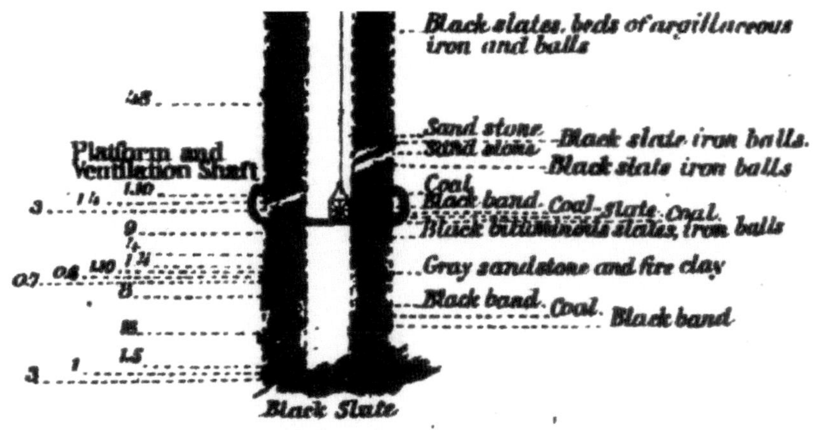

Close-up view of the bottom of the Egypt Coal Mine in 1858. *Author's collection.*

at Egypt, McClane also discovered valuable iron deposits at Buckhorn and devised ways to manufacture the ore at furnaces such as Tysor Place, Endor and Buckhorn.

McClane and Harris set about sinking test pits and boring holes in the earth, searching for coal that the experts reasoned should be present beneath the surface and be of a much better quality than the surface deposits then being worked at the time at several sites across the region. McClane found what they were looking for more than one hundred yards beneath the surface of the earth. In a report on the coal field written in September 1853, Dr. Charles Jackson added a postscript describing what had been found that fall at Egypt: "Since the above report was set up in type, Mr. McClane has discovered a bed of coal, four feet ten inches in thickness, in Egypt, on the south side of the river, where he perforated the coal at a depth of 361 feet from the surface. Our predictions are therefore fulfilled, and the coal has been found at a convenient place for mining."[5]

Shortly after finding the coal seam, Harris decided that the exploitation of valuable coal deposits they had found required more capital than he was able to invest alone. So he sold the mine to a New York corporation known as the Governor's Creek Steam Transportation and Mining Company in 1854. This would prove to be a momentous decision that would lead to several unforeseen consequences in the coming decade.

Utilizing a device described as an "artesian auger," McClane and a team of a dozen miners bored an 8- by 15-foot shaft through clay and sandstone 460 feet into the ground. They used a forty-horsepower steam engine to power the machinery and another to run a pump capable of removing 1 million gallons of water from the mine per day. After describing their initial work of getting the mine in place, a newspaper correspondent noted, "The whole has been accomplished so far without a single loss of life."[6]

Boasts about the safety record of the mine proved to be premature. Within a year, a series of explosions at the Egypt Coal Mine would be the first in a long line of tragic and often fatal accidents that plagued these coal mining activities along the Deep River for nearly one hundred years.

The early deaths in the Egypt Coal Mine were the result of explosions of a substance known as firedamp. Firedamp was a bane to coal miners the world over, and thousands of lives have been lost down through the years thanks to this invisible killer. When describing the various gases miners often encountered in coal mines, Roy Andrew, a mine safety expert who served as the official mine inspector from Ohio in the 1870s, noted that firedamp was the most volatile:

> *The gases generated are mainly proto-carbureted hydrogen gas, or light carbureted hydrogen gas, commonly called firedamp; carbonic acid gas (black damp); carbonic oxide gas (white damp); and sulphureted hydrogen gas. The former of these gases, the fire damp, when mixed with certain properties of atmospheric air, forms a most explosive compound, and is one of the most fatal and dangerous elements encountered in human enterprise. It is emitted from the fissures and minute pores of the coal, and exists in a highly compressed state. It is generally found in greater abundance near dikes and faults in the coal strata, and is more copiously evolved from coals of a highly coking nature than from those of an open burning kind.*[7]

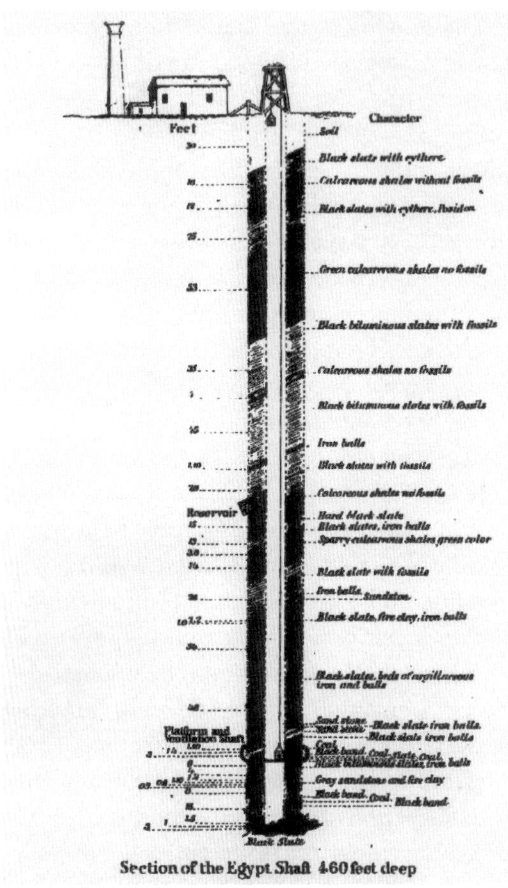

Cross-section of the Egypt Coal Mine shaft, prepared in 1858 as part of Wilkes's report on the mineral resources of the Deep River. *Author's collection.*

The miners at Egypt were aware of the presence of this deadly gas lurking in the depths of the coal mine. In fact, McClane is reported to have treated the gas as some sort of natural spectacle to be used to thrill visitors to the mines. Perhaps he had become dulled to the presence of the gas thanks to frequent contact with the hazard—perhaps familiarity led to lack of respect for the deadly substance.

In the spring of 1856, visitors to the mine were given a vivid demonstration of the presence of this gas when McClane intentionally applied a flame from his "lamp" to an area along the coal seam where firedamp was present. Exactly what type of lamp they were using in the mine at that stage of the operation is unknown. Safety lamps had been developed in Great Britain in the early nineteenth century and had become a common feature in mines on both sides of the Atlantic by the middle of the century. "Mr. McClane touched his lamp to a slight recess in the coal, and we were treated to a specimen of the 'fire-damp,' a beautiful lambent flame, which flickered over the surface of two or three feet, and then gradually went out," observed a writer from the *Fayetteville Observer*. "This fire-damp is a terrible enemy in large mines, where there is any want of care in ventilation. It sometimes produces explosions by which hundreds of lives are lost. But it is easily avoided by proper care. It will be long before there will be any danger in this mine."[8]

Just under a year after McClane was showing off the spectacle of the firedamp dancing along the walls of the mine to his visitors, the first major accident in the Deep River Coal Field occurred when six miners went down into the shaft on Friday evening, February 20, 1857. They were planning to work at a section of the mine just over two hundred feet below the surface. Once they stepped off the mining cart, an explosion of firedamp rocked the mine.

A correspondent from the *Fayetteville Observer* interviewed several witnesses on hand at the time of the blast and collected several interesting details of the incident: "The six men had just descended the shaft, and stepped inside, when the foul air ignited, and its force hurled one man into the shaft, down which he fell and was killed. Another was thrown into a large reservoir of water, and was drowned. Three others were burned to a crisp, mutilated and mangled. While the sixth owes his preservation to a huge spike nail, on which his clothes caught, and prevented his being hurled down the shaft, which is some 200 feet deeper than the vein of coal at which they are now working."[9]

These six miners caught in the blast are of historical significance, as they were the first known fatalities from the Egypt Coal Mine. Unfortunately, their names were not recorded in press accounts of the event. The only personal details we have were noted in a Fayetteville newspaper: "We have not heard their names, but understand that they were all laborers, and Irishmen, and from what we heard of the character of those employed at the mine when we visited last year, we suppose that they were very worthy men."[10]

We do know the name of one of the victims of the blast: John McKeithan. His tombstone survives in the Cumnock Cemetery and notes that he was a native of New Jersey. The other victims of the blast might be buried in unmarked graves in the cemetery, but so far, McKeithan's is the earliest identified.

Work quickly resumed at the Egypt Coal Mine, and nearly two months passed without incident. Extra precautions were taken, including opening a ventilation shaft that the owners hoped would provide adequate fresh air and alleviate danger of another firedamp explosion. Then, on the morning of Monday, April 6, 1857, another explosion ripped through the Egypt Coal Mine. This blast took the lives of five miners: Robert Dunn, George Lewis, Michael McCormick, Daniel Hayes and J. Byrne. Once again, it appears that firedamp was to blame.[11]

A letter written shortly after the explosion described what happened: "The explosion took place at 8 o'clock, a.m., on Monday last, at the Governor's Creek Coal Mines, which are situated at a place in Chatham county [*sic*] known as Egypt. The Superintendent, Robert Dunn, and four of the hands, all white men, were instantly killed, and two others were thought to be dead; but after being buried awhile, (with apertures in the ground giving their nose and eyes fresh air—the method adopted, I understand, for the purpose of relieving the lungs of the sufferers from foul air) they were restored to life; but the life of one of them, a man named Carter, is despaired of. The damage to the property is said to be very great."[12]

The dust had hardly settled from the blast when three brave miners volunteered to descend into the shaft to retrieve the bodies of the dead. They nearly became casualties themselves when they were overcome by gas when they reached the 430-foot level. One of the men passed out from the effects of the gas. His two companions were also affected by the fumes, but they managed to tie a rope around themselves and signal the men on the surface to pull them out. They lost consciousness before reaching the surface and were out of commission for several minutes. Once their heads were cleared up a bit, they returned into the mine to bring out their companion, who still lay unconscious at the bottom of the mine. Fortunately, their friend was still alive when they got to him, and they were able to hoist him out of the mine.

Miners descended into the shaft on April 7 to retrieve the bodies of the dead and to investigate the damage. After sending the bodies to the surface, they looked around inside the mine. They found that most of the recently installed machinery had been destroyed in the blast, including the water pumps. Without the pumps, water quickly flooded the shaft, filling up the mine until leveling off at the two-hundred-foot level.[13]

The loss of the superintendent was a big blow to the fledgling mining operations. Robert Dunn, a native of Scotland, was apparently brought in by McClane for his mining expertise. Whether the two had worked together in Pennsylvania at some point prior to their work at Egypt is unknown. Dunn was well liked by members of the Egypt community. He left behind a widow and five children. An unnamed individual wrote of Dunn, "Those who have visited the Egypt mine will remember the kindly face of Mr. Dunn, who, under McClane, was the superintendent of the works, and whose politeness to visitors rendered him a favorite. We heard of his sad fate with great regret."[14]

There was at least one other Scottish miner among the dead. Thirty-eight-year-old Michael McCormick was a native of Dumbartonshire. Whether he and Dunn were acquainted with each other in the old country is unknown. There was also one Irishman among the dead: Daniel Hayes, who hailed from Limerick. His tombstone also tells us that he was twenty-four years old at the time of the accident. Hayes and McCormick were interred at

Cumnock Cemetery near the grave of at least one of the victims from the previous mine explosion. It is uncertain where Lewis and Byrne were buried. They may have been put to rest at Cumnock in unmarked graves, or their headstones may have been lost down through the years.

During the fall of 1857, an interested individual traveled overland from Fayetteville to Egypt to see firsthand how work was going at the coal mine. He hunted down Superintendent McClane to question him about the status of the mine. McClane was very forthcoming, stating that the operations were completely shut down at the moment but that the mine could be put back into production with sixteen days' notice. He also noted that the miners at Egypt were capable of producing coal at the rate of one ton per minute.[15]

McClane then gave the inquisitive visitor a grand tour of the place, showing him some of the extensive work that had already been completed at Egypt. Although the shaft was nearly full of water and they were thus unable to descend into the mine, they did get a close-up view of much of the related machinery. The man from Fayetteville noted:

> *Connected with the shaft by rods and other apparatus are two splendid steam engines, each of 60 horse power, which are well worthy a more deliberate examination than I could bestow. Besides the large building which encloses these engines, I noticed others used as saw-mills, blacksmith shops, etc. The mouth of the shaft is covered with a scaffolding which supports the wheel around which a rope revolves attached to a bucket or cage, which traverses the mine, and is drawn up or let down as occasion demands.*
>
> *The engines and chimney rest upon bases of solid hewn sandstone rock, quarried in the vicinity and have a very substantial appearance. The chimney is about 80 feet high. From the mouth of the mine to the river, a distance of 300 yards, a small railroad is already constructed, and as the land shelves in that direction, the grade is overcome by trestle-work about 30 feet high, in some places. The trestle-work is filled in with a curious looking slate rock, which, upon inquiry, I ascertained was not less than the celebrated iron ore, discovered by Mr. Paton in 1856, to be identical with that from which "pig-iron" is made in Scotland. Prof. Emmons alludes to this discovery in his last report. Before its value was discovered, some $30,000 worth had been expended in filling in the trestle-work.*
>
> *The river is about 180 feet wide, and some 6 feet deep at Egypt; a canal is already begun to conduct the river under the superstructure of the road, so as to "dump" the mineral at once into the boats from the cars above.*[16]

Despite the tragic episodes of 1857, work to develop the mineral resources of the Deep River Coal Field continued. Other locations for coal mines were explored, especially in the vicinity of Gulf. In addition to the coal, efforts were underway to exploit the other valuable minerals, including iron and copper. Gold was even found in small quantities nearby.

A major impediment to progress was transportation of the resources to market. Most looked to the rivers of the area as their best hope for an economical outlet for the commodities of the region, both industrial and agricultural. But several navigation companies had failed to make the Cape Fear and Deep Rivers navigable on a reliable basis. As early as 1792, companies such as the Deep and Haw River Navigation Company, the Cape Fear Company and the Cape Fear Navigation Company had tried without success to figure out how to overcome the rocks and rapids in the river upstream from Fayetteville. Although the river was open for batteaux and steamboats on a few occasions, navigation was sporadic above Fayetteville for all but the timber-rafters in the first half of the nineteenth century.

In 1849, a new company, the Cape Fear & Deep River Navigation Company, took up the challenge. It hoped to provide slack water navigation from Fayetteville upstream to Waddell's Ferry on the Deep River in Randolph County. In addition to this ambitious goal, it also planned to connect the Cape Fear and Yadkin River systems by means of a portage railroad.

One of the main factors spurring development of the navigation project was the potential financial windfall expected once the mineral resources of the Deep River region were fully developed. Promoters of the region dreamed of the day when steamboats towing barges filled with coal and copper would make their way in a long and steady line down the Cape Fear to Wilmington, and from there these natural resources would be sent to markets throughout the world. Unfortunately for these visionaries, the project never lived up to its potential. Although the Cape Fear & Deep River Navigation Company was the most successful of the navigation companies to take on the challenges of the upper Cape Fear, it was never able to overcome the obstacles encountered at places such as Smiley's Falls and Red Rock.[17]

The Cape Fear & Deep River Navigation Company pared back its initial enthusiasm and set the more attainable goal of opening the rivers as far upstream as Hancock's Mill on the Deep River in Moore County, where valuable deposits of coal and pyrophyllite were discovered. By the middle of the 1850s, the main obstacle blocking the opening of the river system was a series of rapids along the Deep River at Lockville. Locks and dams upstream on the Deep River were in place, as were those on the Cape Fear, although the latter were frequently out of commission due to floods and freshets.

By 1860, with rumors of war stirring in the air, the company received an infusion of capital from the State of North Carolina, which took over direct management of the project. Ellwood Morris was appointed by Governor John W. Ellis as the engineer assigned to oversee work on the rivers. Morris's crews pushed the work forward, and by 1859, they had the rivers open all the way to the coal field at Egypt and Gulf.

As he became acquainted with various people involved with the navigation efforts, as well as the development of the region's mineral resources, Morris became disenchanted with the lack of effort on the part of the mine owners to extract coal for transport along the river. By the summer of 1860, with the navigation works in good working order all along the line, there was no coal waiting for shipment to Wilmington. Morris reported that he could have passed coal boats all the way through the works by September 1 without interruption, but there were no boatloads of coal.

While aboard the company's steamboat, *Haughton*, moored near Averasboro in Harnett County on September 3, 1860, Morris noted some of his concerns in a letter to Governor Ellis:

> *So far as I hear from* the Coal Field Men, *they are making no preparations to ship coal,* in quantity, *either by navigation, or by Railroad—& I am reluctantly reaching the conclusion, that the entire coal field is in the hands of* mere speculators, *whose object is to realize heavy profits—not by* working *their lands, but by* selling them.
>
> *"If this view should prove correct, (and I* hope *it may not) the Coal lands must change hands, before the coal basin is developed, & hence the future both of Navigation & Railroad, will remain for some years longer uncertain, as without* coal in large quantity, *neither can ever pay.*[18]

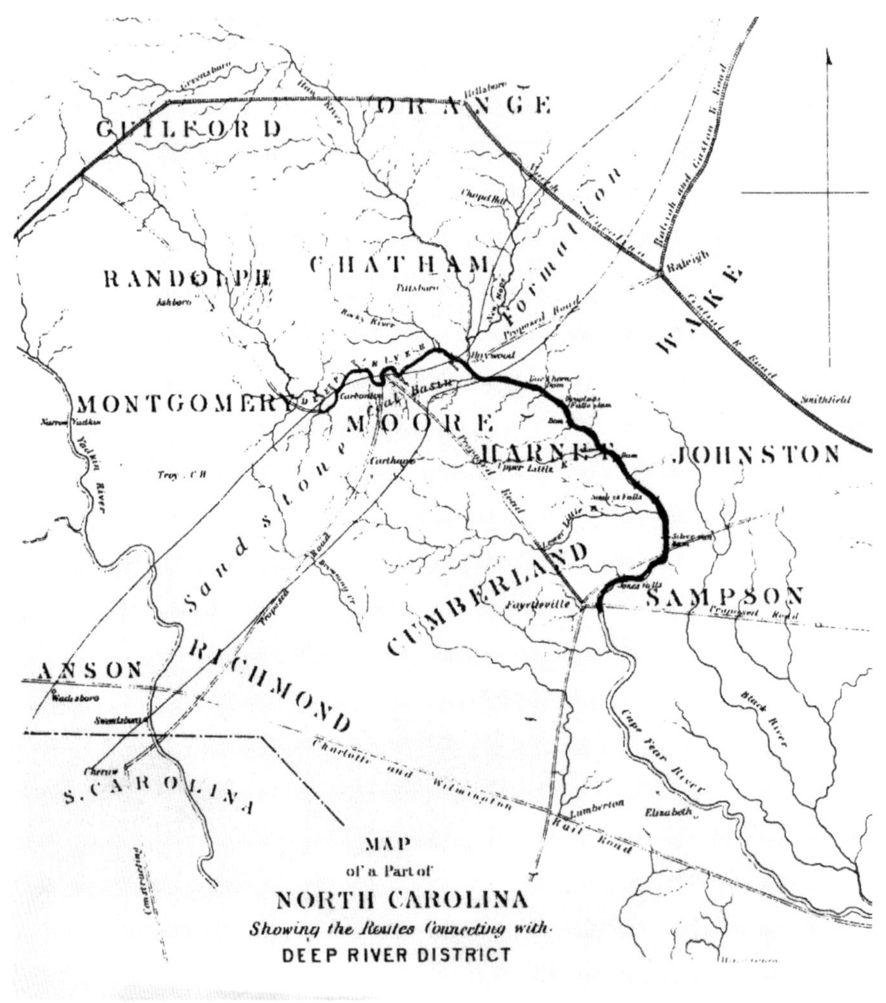

This map of the Deep River Coal Field was published in 1860. *Map courtesy North Carolina Archives and History.*

Chapter 3

CIVIL WAR COAL MINING DISASTERS

With the onset of the Civil War, activity in the Deep River Coal Field increased to a frenzy as the people prepared for war. New coal mines and iron furnaces were planned to deliver the raw materials that the Southerners would need if they wanted to have a chance of winning their independence.

The Egypt Mine was no exception. Mine officials scrambled to remove water from the flooded shaft and repair machinery damaged by the explosions in 1857. They also worked to deal with many of the transportation problems that hampered their efforts to get the coal to the outside world. As the 1860s began, the terminus of the Western Railroad at McIver's Depot was nearly three miles over a rough cart path from the Egypt Coal Mine, while the improvements of the Cape Fear & Deep River Navigation Company were out of commission downstream from Lillington.

There were many deposits of coal, iron and other minerals in the Deep River country, and with the onset of war, there was a dire need to get these valuable and much-needed resources to places where they could be turned into the raw material to support the war effort. The state lacked the manpower to keep the workforce employed on the Cape Fear to repair the locks and dams downstream from Northington's Falls that had been damaged by forces of nature as well as vandalism. But the minerals of the region, especially the iron ore at Buckhorn and the coal at Egypt, were too important to the Confederate war effort for officials to allow them to remain underground untapped.

To overcome the problems of transportation, the state decided to keep the locks and dams upstream from Battles Lock and Dam to Gulf open to allow transport on the river to Egypt. There the cargo could be placed on the Western Railroad and shipped overland to Fayetteville, where it could then be put back on a boat to continue the trip downstream to Wilmington. The railroad ran from Fayetteville just to the edge of the coal field, where the company was forced to stop its line's progress thanks to a rock outcropping just north of McIver's Depot. With the war approaching, company officials redoubled their efforts and pushed the railroad north to the Egypt Coal Mine.

While other mines in the area were being brought into production, work at the Egypt Mine was slowed over legal issues surrounding the status of the mine's owners, who were predominantly from the North. Thus, when the war broke out, what at the time was the region's best-known coal mining operation lay idle.

Work had resumed at the Egypt Mine by the fall of 1861. The first load of Egypt coal made its way over a two-and-a-half-mile cart path to McIver's Depot, where it was transferred from the carts and placed into the railroad cars of the Western Railroad. From McIver's it was taken to Fayetteville and from there via steamer to Wilmington. This first load of coal for the Confederate war effort was earmarked for Charleston, South Carolina.

One newspaper editor complimented McClane on his perseverance in the face of the numerous setbacks he had encountered in his attempts to exploit the coal and other mineral resources of the Deep River: "The interruptions to the business have been unfortunate, but we trust they are now overcome, and that Mr. McClane, with his accustomed energy, which has sustained him through all the disappointments about Deep River navigation and of fire-damp explosions, will now begin to realize the full and profitable results of his long and skillful labors unearthing the wealth of that wonderful region of our State. In these labors he has persevered for years, and deserves not only pecuniary benefits but a grateful appreciation by those who are interested in the prosperity of the State."[19]

In November 1861, A.C. Murdock paid a visit to Egypt to see how work was progressing. He descended 455 feet into the Egypt Coal Mine and returned to the surface with a piece of the coal, as well as a sample of the black-band iron ore. He promptly forwarded these to the editor of the *Hillsborough Recorder*, with a note extolling the virtues of the mineral resources of the area: "They are now taking out coal in abundance, making iron in small way, and also making oil. I have no doubt but the coalfields of Chatham, will soon be the great Lowell of the Confederate States."[20]

Coal was also being produced from nearby mines at a surprising rate. John Colville, who was in charge of the coal mining operations at the Taylor Place, shipped seventy tons of coal down the railroad to Fayetteville in the first week of December 1861. Colville was also bringing other mines into operation as quickly as possible: "And we further learn that the Haughton property, adjoining the Taylor place, is to be worked forthwith, also under the superintendence of Mr. Colville. Mr. Colville has also, as we hear, taken out some very superior Coal from the place lately belonging to the heirs of McIver, only about half a mile from the present terminus of the railroad."[21]

Work at the Egypt Coal Mine was suspended in 1862—not because of an accident but because of a legal technicality. Back in 1854, Brooks Harris sold the mine to a company based in New York. From that time forward, the owners of the mine were predominantly Northerners and were viewed by many as merely speculators more interested in turning a quick profit from shady deals than actually putting together a vibrant working mining operation that would benefit the members of the community. These were not just xenophobic notions on the part of the locals because the owners were Northerners. New Jersey resident Ellwood Morris, the engineer assigned to work on the navigation works by Governor Ellis, voiced his concerns prior to the war about the owners of the coal mines being mere speculators. When the war started, the fact that the mines were owned by Northerners put the company in a state of legal limbo. The Confederacy needed the coal, but there was the matter of clearing up these absentee owners before large investments could be made in the mine.

At the November term of the Confederate Court for the District of North Carolina, James Browne of Charleston and Charles B. Mallett of Fayetteville were appointed to oversee operations at the mine. A notice published in several papers throughout the region pointed out that the two men had "entered into a partnership for the purpose of mining and selling Coal, and solicit orders for the same in any desired quantity. Orders for any amount can be supplied on short notice. The coal from their property is undoubtedly the best in the Confederate States."[22]

Suspension of the work was blamed on a lack of rope, but this was just a cover story. Once the legal matters were taken care of, work at the mine continued: "We are pleased to state that after a suspension of operations for some time for want of a rope, the working of this mine has been resumed, under other management than formerly. We believe that it has been placed by the Confederate Court in the charge of C.B. Mallett, Esq, President of the Coalfields Rail Road, under whose management no one doubts that it will be worked to the best possible advantage."[23]

Work at the Egypt Coal Mine continued in earnest, with the miners bringing up a steady supply of the black rocks. Despite their success, the mine developed a sinister reputation. We know that there were accidents in the Egypt Coal Mine during the Civil War period, but the extent of the accidents is uncertain. Mine officials and Confederate authorities kept a tight lid on news coming out of the mine, especially news that might scare off any potential miners.

There are vague accounts of mining accidents during the period that hint at some rather significant episodes with multiple fatalities. One explosion is rumored to have claimed the lives of more than fifty people. Unfortunately, contemporary eyewitness accounts of this disaster have not come to light, even though tales of the tragedy made its way into the local lore.

State Geologist J.A. Holmes was familiar with the Civil War–era disasters, apparently getting information firsthand from those who lived in the area during the war years. Back in 1890, while visiting the mine with a group of students, he gave a hint at the calamities that had rocked the Egypt Mine twenty years earlier: "The mine was operated for several years prior to the beginning of the war, whether with financial success or not I am unable to say. During the war it was operated with varying success. The methods of mining were crude, the ventilation of the mine was poor, and times the FIRE DAMP [gas] allowed to accumulate in the mine, exploded, killing or injuring badly every man in the mine."

At some point during the war, miners from Cornwall, England, were employed to work at the Egypt Coal Mine. Cornish miners had made their way into North Carolina by the 1830s, lured mainly into the gold and copper mines of the Piedmont. Their knowledge and technological skills proved a valuable contribution to the state's gold mining operations, especially so after the mines were sunk deep underground as lode mining expanded.[24] There is little wonder that the miners from Cornwall found their way into the coal mines along the Deep River, especially the subterranean operations at Egypt.

How early they were working at Egypt is unknown, as is the extent of it. Most surviving accounts of the early days of the mine mention miners coming mainly from Scotland and Ireland. Tombstones in the cemetery at Cumnock demonstrate that men from Cornwall were present at the Egypt Coal Mine during the early 1860s.

The two Cornish miners were killed in an explosion in the Egypt Coal Mine on November 10, 1863. This is the best documented mining disaster at Egypt during the Civil War period. An explosion ripped through the mine at 10:30 p.m. on Tuesday, November 10, 1863, taking the lives of six miners. Word of the disaster leaked out, and brief notes appeared in some of the newspapers. It is interesting to note that company officials were eager to downplay the possibility that the explosion was caused by gas and pointed out that the mine had more than adequate ventilation to prevent such an accident.

One contemporary account of the accident gives some interesting details of the disaster: "We learn that it is supposed that the disastrous explosion in the Egypt coal mine on Tuesday night was of powder used for blasting and not of fire damp. This impression is derived from statements made by one of the six men who lost their lives. He was taken out alive and before he died gave some particulars that indicated the absence of fire damp. The excellent arrangements for ventilation lead to the same conclusion. It is expected that the works will be in operation again."[25]

Other press accounts of the accident provide interesting details about the men employed in the mine at that stage of the war: "We regret to learn that on Tuesday night last at 10½ o'clock an explosion occurred in the Egypt coal mine, by which six men were killed and the machinery damaged. We did not learn the names of the men, but it is said that five of them were conscripts, who had been detached to work in the mine to escape service in the army."[26]

As stated earlier, gravestones in the cemetery at Cumnock indicate that at least two of the men killed in this explosion were Cornish miners from England. Solomon Stone's tombstone states that he was fifty-one years old at the time of his death and a native of Cornwall, England. Alexander B. Warren's tombstone notes that he was from Cornwall and was thirty-nine years old when killed in the blast.

There is approximately twenty feet of open space between the graves of the two Cornish miners. It is possible that the other victims of this accident are buried in the intervening space, but that is uncertain at the present time. Advanced archaeological techniques such as ground-penetrating radar can be brought in at some point in the future to verify the existence of graves in this open space and can verify whether or not bodies are present. These techniques will unfortunately not be able to reveal the identity of those interred.

The Confederacy could ill afford to let one of its most important sources of coal remain idle. Repairs were made to the damaged machinery, and work resumed.

Another group of miners employed in the Egypt Coal Mine during the war was a group of Union army deserters who found themselves in limbo after being paroled by the Confederate government and released under their own recognizance in Richmond in the winter of 1863–64. The people of Richmond were having problems with these quasi-prisoners congregating about the city, many of whom were causing trouble with the local populace. The mayor of Richmond, Joseph Mayo, appealed to President Davis for help but was unsuccessful in obtaining relief. Finally, a "Mr. Baxter" came up with a solution to put these men to work in the coal mines. "Mr. Baxter, a Government official, was preparing a scheme to send these deserters to coal mines or somewhere else. He, however, did not think it would work."[27]

Councilman Burr pointed out, "We ought to object utterly to the paroling here at least of any genuine Yankees. He would not object to their paroling Irishmen. But the genuine Yankee should not be turned loose here."[28]

We do not know why Baxter did not have full confidence in his plan, but the former Union soldiers were becoming such a nuisance that something had to be done. Within a week of the discussion with the mayor and city council, the first group of prisoners had been rounded up and sent to work in the coal mines of central North Carolina. A correspondent for the *Richmond Whig* noted, "Yesterday morning, forty-three Yankee deserters were, with their own consent, sent to Wilmington, North Carolina, to work in the coal mines near that place. This is a decided improvement upon the plan, until recently practiced by the Confederate authorities, of turning these creatures loose in the city to prey upon the honest and industrious portion of the inhabitants."[29]

News of the experiment spread south, where a columnist for the *Charleston Mercury* added a few details about the workforce: "The local columns of the Richmond *Dispatch* state that forty-three Yankee deserters confined in Castle Thunder were to be sent off that morning to Wilmington, N.C. It is the purpose to place them in the coal mines, and thereby make them useful to the Confederacy."[30]

The laborers made their way to the Deep River country in a roundabout fashion. They were sent to Wilmington on the Wilmington and Weldon Railroad and then put on steamboats and shipped up the Cape Fear River to Fayetteville, where they found the rail cars of the Western Railroad ready to take them to Egypt.

After arriving at the mines in Chatham County, many of the men became disenchanted with the turn of events, and some tried to escape. "Seven of the Yankee deserters re-deserted on the night of their arrival at the coal mines in North Carolina. Five have been recaptured."[31] What punishments these men received for trying to desert their newfound vocation is not recorded. In all likelihood they were returned to work in the mines.

According to information compiled by Guy Smith (via personal communication), by May 1864 more than one hundred men, including forty Union prisoners, were hard at work underground at Egypt. Their peak output occurred that same month when miners brought out fifty-eight tons of coal in a single day.

The true scope of the coal mining operations in the Deep River Coal Field during the Civil War period is not well understood, and the importance of the mines to the Confederate war effort is often underestimated by historians and geologists alike. According to records compiled by the U.S. Geological Survey, coal production in North Carolina was at its highest point of nineteenth-century production during the war years, with 30,000 short tons produced in 1862, 30,000 short tons in 1863, 25,000 short tons in 1864 and 20,000 short tons in 1865. This compares with peak postwar production during the rest of the century, which was 26,896 short tons in 1899.[32]

Few records of the wartime coal operations survive. We do know that in addition to the Egypt Coal Mine there were several other mines in the area being worked, including mines at Gulf and Farmville. One of the most enterprising of the individuals involved in the Deep River mineral operations was the aforementioned John Colville (or Colvin, as his name is sometimes transcribed). Colville was a man of many talents and even obtained a patent for some of his metallurgical work down at Wilmington prior to the war. He is best remembered for his work at Buckhorn, where he devised the ironworks there that included a gravity-propelled tramway and log-pen blast furnace. The blast furnace, built in haste under the exigencies of wartime supply shortages, was found to be more efficient in operation than nearby furnaces that were built using state-of-the-art technology. The Ocknock Furnace was destroyed by General Sherman's troops at the end of the war, but remains of the tramway he devised, as well as the modifications to the old canal, are still visible.

Most people are unfamiliar with Colville's work in the coal operations. However, he was very industrious in his efforts to develop the mineral resources in the region to meet growing wartime needs. One of the discoveries he made was a vein of coal described as "semi-bituminous," from an as-yet-unidentified location along the Deep River:

> *We were surprised to learn from a friend who was recently in the mineral region, that Mr. Colville has within a few weeks past opened a shaft to the*

> *depth of upwards of 200 feet into this deposit of anthracite, and that he is mining it with his accustomed energy. At present we learn that this coal is higher priced than the bituminous was, (when it could be had) owing to the distance it has to be hauled. When the railroad shall be completed to that part of the coal fields, its cost will doubtless be reduced one-third or one-half. It is still, however, not more expensive than the bituminous, as it burns much longer, a good fire requiring to be renewed not oftener than once in the course of a day. And, by the way, in rooms where a very hot fire is not required, a single fire a day may suffice, by making a sort of paste of the coal ashes and spreading it over the top of the fire after it is thoroughly ignited, taking care to leave an open space at the back for the escape of gas.*[33]

There are a few incomplete records of the Mallett and Brown operations at Egypt that hint at the extent of the coal production at their mine. During July 1864, their agent in Wilmington reported selling various consignments of coal totaling more than 575 tons of coal for sales totaling $32,990. This included an 85-ton purchase for the steamer *Florrie*. Earlier that year, on May 15 and 16, they sold two flatboat loads of coal shipped down the river totaling 152 tons.[34]

Work at the Egypt Coal Mine continued right up to the end of the war. As the Union's grip on the beleaguered Confederacy tightened, the mineral resources along the Deep River became even more important to the war effort. The Deep River Coal Field was reputed to be the last remaining source of coal for the Confederacy in the final months of the war.

The most famous wartime episode at Egypt was not a mining accident. Egypt gained some note of fame as the hiding place for the machinery from the Fayetteville Arsenal. As the forces of Union general William T. Sherman approached Fayetteville from the southwest, Confederate defenders sent the valuable machinery up the Western Railroad for safekeeping. This proved only a temporary reprieve, as Sherman's forces nearly caught up with the machinery in April 1865.

With General Joseph E. Johnston's surrender imminent and Union military forces bearing down on the Deep River coal operations, desperate Confederates decided to destroy the machinery instead of letting it fall into enemy hands. They lowered the machinery deep into the Egypt shaft and then set off an explosive charge that destroyed much of the machinery and sealed off the side tunnel.

Union forces occupying Raleigh in the final days of the war heard rumors about the machinery from the Fayetteville Arsenal being hidden somewhere along the Deep River. Members of Forty-Seventh New York Infantry Regiment were dispatched on May 9, 1865, to look for the machinery. They found what they were looking for in the Egypt Coal Mine and spent three days gathering together what was left of this important machinery that had originally been captured at Harpers Ferry in 1862. The official report from Company D gives interesting details of the mission: "We crossed the deep river near Lockville on a bridge the same day and camped eight miles from Egypt. On May 9 we loaded the wagons with machinery that had been stolen from Harper's Ferry. We remained three days. Loading the wagons and then we commenced our return on May 13 and reached Raleigh on May 16, where we are at present encamped."[35]

The destruction of the Fayetteville Arsenal machinery did not totally end operations at the Egypt Coal Mine. By the spring of 1866, the mine was open under the management of Robert Paton. By 1870, work had come to a halt, and the mine lay idle for more than a decade. Plenty of coal remained underground, but the war had taken its toll on the people of the Deep River country.

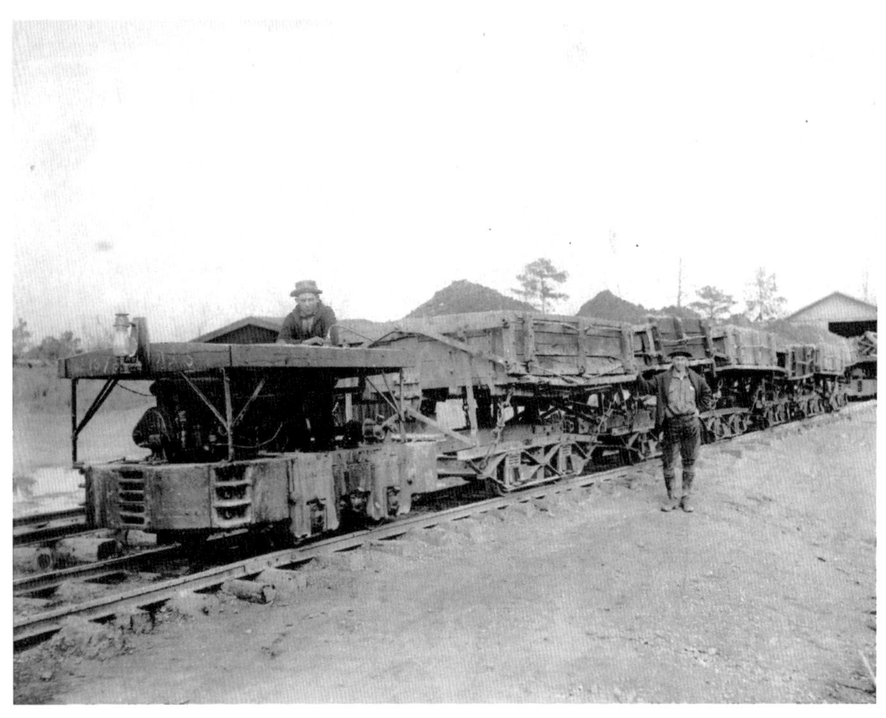

A load of coal from the Cumnock Mine, circa 1900. *Courtesy Guy Smith.*

Chapter 4

THE EGYPT COAL MINE BECOMES THE CUMNOCK COAL MINE

Henry M. Chance made a visit to the various coal-producing locations in North Carolina in the early 1880s. He visited the Egypt Coal Mine in 1884 and found the place in a state of disrepair. "At present the colliery is a total wreck," observed Chance. "The shaft was caved in and the head-frame rotted and fallen into the shaft; the boilers are partly dismantled, the engines badly rusted, and the great brick smoke-stack is almost ready to fall, if indeed it has not fallen before this is printed, the lower courses of masonry being built of a light colored sandstone that has disintegrated rapidly so that it can now be broken off and crushed to sand in the palm of the hand."[36]

Not long after Chance's report on the coal resources of North Carolina was published, a group of northern industrialists led by a man named Samuel Alexander Henszey came to the Deep River country and purchased the Egypt Coal Mine. Whether they were lured by Chance's glowing report is unknown, but in 1888, this group, operating under the name of the Egypt Coal Company, started operations in Chatham County. This company was initially successful in its work at Egypt.[37]

Henszey was a native of Philadelphia, Pennsylvania, who became involved in the operation of railroads and mining at an early age. He was working as vice-president and general manager of the Arizona Central Railroad before coming to North Carolina in 1887. He was lured to the state by the economic potential offered from the development of the Egypt Coal Mine, as well as railroad facilities across the central portion of the state.

Shortly after he organized the Raleigh and Western Railroad, a contemporary wrote of Henszey, "Thus he became the controlling head of a profitable railroad and of a coal-mine of exceptional richness and value, far removed from any competitor. The mine provides the railroad with sufficient freight to make it pay handsomely and independently of any other patronage, while the railroad enables the output of the mine to be marketed to the best possible advantage."[38]

The first order of business at Egypt was clearing the mine of water and debris. There are few details about how much damaged and twisted machinery remained from the Fayetteville Arsenal after the Union strike force visited the area in May 1865, and we do not know the extent of damaged machinery from the coal operations of Mallett and Browne, which was blown up at the end of the war. Before they could get to this, Henszey and his partners faced the challenge of clearing the mine of water. They spent four months continuously pumping water twenty-four hours a day, seven days a week, at a rate of six hundred gallons per minute.[39]

Once it was free of water, workmen descended into the mine to survey the extent of the damage and to figure out the next steps in the process of opening the mine: "The shaft of the mine (which is 7 by 14 feet) was sunk to a depth of 466 feet, and several tunnels were run a considerable distance, one of them being 440 yards long."[40]

In addition to restoring the mine, the developers had grand plans for building a thriving town at Egypt. This would be something not just for the miners and their families but also for tourists whom they hoped to attract to the property. They laid out the new town on a 125-acre tract of land just east of the public road leading to the covered bridge across the Deep River. A key part of the planned resort was the Egyptian Inn, a wooden-shingled Queen Anne–style building they built on a high piece of land near the train depot.

The most widely reported event happening at the mine was not the community development activities. Instead, it was an accident underground in the Egypt Coal Mine that grabbed the headlines on front pages of newspapers across the country.

On the afternoon of February 20, 1890, a rock slide in the main shaft caught the access cage midway between the surface and the bottom of the mine, between 200 and 250 feet below the surface, completely sealing off the mine from above. In addition to depriving the miners of their only way out of the mine, the rock slide also had several dangerous consequences. The track and cage were wedged so thoroughly as to cut off light and air entering from above. Communication between the miners was knocked out. No sounds of people on the surface could

Signs of Petroleum.

Capt. Smith, of the C. F. & Y. V. steam ferry boat Compton, states that the presence of oil in the Egypt coal now being burned by that vessel, is so manifest that coal piled on the deck of the vessel while being put into the hold, leaves a greasy place that is quite perceptible.

This reminds us that in his report concerning the Egypt coal fields Prof. Emmons says there are also bituminous slates connected with this coal. The professor says: "From 30 to 40 gallons of crude kerosene oil exists in every ton of these slates. They are from 50 to 70 feet thick, and it is proper to state that it is a better oil than is furnished from coal."

The land in the vicinity of the Egypt Coal Mine has long been reputed to contain many valuable commodities, as this article from the *Wilmington Messenger* of March 12, 1890, explains. *Author's collection.*

be heard by the men in the mine. At the same time, rescuers heard nothing but silence in response to their cries and hallows into the mine.

Asphyxiation was not a concern for the miners, thanks to the ventilation shaft dug by McClane and his men back in 1857. The most ominous threat was the fact that the power to the water pumps was knocked out. This could prove fatal to the miners below, as the Egypt Mine was known to fill quickly with groundwater. The men on the surface worked through the night trying to reach their trapped companions. A head count taken shortly after the accident occurred showed that at least forty miners were down below.

Rescuers cleared rock and debris from the shaft. Finally, just as the sun was rising, they were able to cut a hole in the wall of the cage, thus allowing light and air into the mine. They immediately called out into the mine, and much to their relief, answers were returned from below. Ropes were lowered, and the trapped miners were hauled to safety:[41]

> *We did not expect to be reached at all....We felt sure that the mine had caved in at the top and not a man of us ever expected to be taken out of that pit alive. We huddled close together and spent the time singing and praying. We knew at the rate water was rising on us that it would be only a matter of a few hours before we would all drown, and it required a lot of talk and persuasion to keep some of the men from lying down and drowning before*

> *the water was three feet deep. We then made a venture to stand on our feet just as long as we could and when we could stand no longer we had agreed to all lay down in the water at the same time and die. It was an awful time and I think we all suffered the horrors of a hundred deaths.*[42]

Despite the setback and negative publicity generated by the near fatal accident in the winter of 1890, improvements continued to be made to the Egypt Mine as more money from investors flowed in. Company president Henszey explained about his company's work, "The coal outlook at our mines is better than ever before. We have nearly completed and will have in use within sixty days a hoisting engine with a capacity of 1,000 tons in ten hours. The quality and quantity of the coal improve as we go down. The depth of the mine now is 800 feet. Competent mining engineers, after long and careful examination, say that we have 12,000,000 tons of coal. That is based on present openings and specific gravity of coal, as furnished by State Chemist Battle."[43]

The company changed the name of the town and mine from Egypt to Cumnock in 1895. The name was officially changed on May 11, 1895.[44] The reason for the change is uncertain, but the most likely theory is that the company was having difficulty recruiting miners because of the dangerous reputation of the Egypt Mine thanks to nearly forty years of fatal accidents at the mine. Others have suggested that the name was changed to honor a company official, but so far, no company official named Cumnock has been identified.

Another theory behind the name change suggests that the town and mine were named in honor of a town in Ayrshire, Scotland, called Cumnock. This was a town with a long tradition of mining. In fact, touring guidebooks issued as recently as the mid-twentieth century referred to the Scottish Cumnock as "a grey mining town."[45] Scottish miners had been working at the Egypt Coal Mine since the earliest days of its existence, and many people in the community had deep Scottish roots, so it is not inconceivable that they would name the settlement for a town in Scotland. Regardless of where the name came from, the new name was used mainly by outsiders, with locals preferring to stick with the old name of Egypt.

News of the new name spread across the region. Press reports noted, "The coal mining industry in this State is at last developed. There are 250 men at work in the Cumnock mines, in Chatham county. The mines used to be the Egypt mines, but last week the name was changed."[46]

Governor Elias Carr visited Cumnock in the fall of 1895 and descended into the mine to personally investigate the workings. In addition to being a leader of rural development, he was a strong advocate of developing

North Carolina's mineral resources. While being shown the coal seams underground, he noticed that when workmen held up their lantern near the wall of the mine to provide illumination, small discharges or explosions of gas would occur. Carr observed that the miners were using regular lanterns with an open flame instead of safety lanterns. As events played out, this was an omen of a tragic event soon to follow the governor's visit to the Cumnock mine.[47]

On December 19, 1895, about seventy miners descended underground to work in Slopes 1, 2 and 3 at the Egypt Coal Mine, 460 feet below the surface. The day started as just a routine day, but at about 9:00 a.m., an ominous noise that many later likened to the deep rumble of an earthquake was heard and felt all across Cumnock and the surrounding community. The people who felt it knew that something tragic had just happened in the mine.

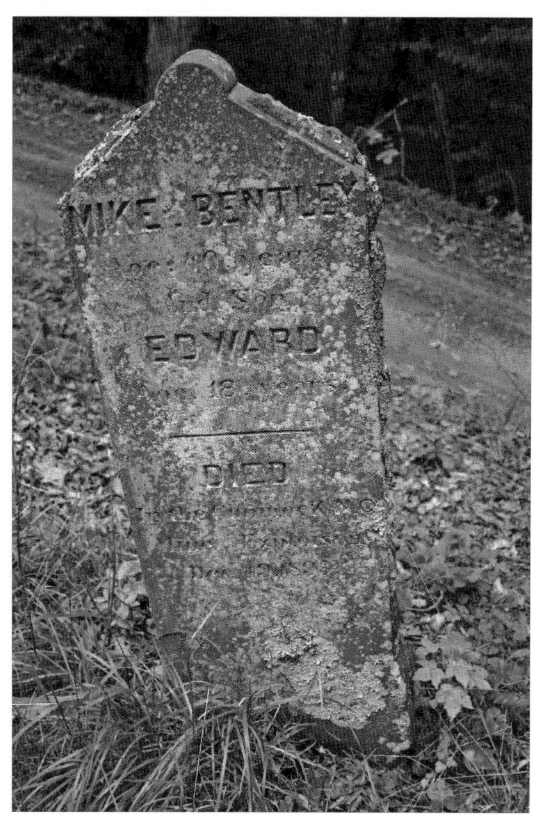

Tombstone at Cumnock Cemetery of father-and-son team Mike Bentley and Edward Bentley, who were killed in the Cumnock explosion in December 1895. *Photo by author.*

Families of the miners rushed to the mine opening to see what had happened. There they found men on the surface trying frantically to contact the miners underground. For more than an hour, no signs of life came from inside the mine. Rescuers then began pumping air into the mine in hopes of providing oxygen for any trapped survivors, as well as dispersing any dangerous gases that might still be present. Their efforts were rewarded when men began emerging from Slopes 2 and 3. Sadly, only about one-third of the men working in the mine were working on those slopes. The rest were on Slope 1, which remained eerily silent.

A group of about a dozen miners volunteered to descend into the shaft and investigate why no one had heard from the men on Slope 1. They descended into the mine, where they discovered that the gas was still strong—so strong that many members of the rescue team were overcome by the gas and began coughing up blood. The rescuers pressed on. When they finally reached Slope 1, the passage was strewn with debris, including twisted iron rails, mangled coal cars and broken timbers. This debris did not completely block their path.

Upon arriving at the spot where the miners had been working, they found more signs of a terrible explosion, with torn and mangled bodies strewn about the mine. This group of rescuers immediately returned to the surface to report their findings to the superintendent.

Another rescue party was assembled to descend into the shaft and bring back the remains of the dead. The recovery effort dragged on into the night. Finally, by about 1:00 a.m., men bearing the first bodies made their way to the surface. A correspondent from the local newspaper was present at Cumnock when these first bodies were brought up and interviewed one of the members of the rescue team afterward about what he had seen deep below the earth: "One of the men who helped to rescue the bodies informs the reporter that he found dead men with their brains and eyes knocked out. Some of them had their shoes blown off their feet. Some of them were carried on stretchers up an incline of nine-hundred feet."[48]

The disaster at Cumnock was national news, and word of the event spread across the country. Even the *New York Times* carried a brief note about the calamity on the day following the explosion: "The explosion was heard at Moncure, fourteen miles distant. Physicians have been sent to Cumnock from all points in that section, and special trains are heading to the mines."[49]

A few months after this particular mining disaster, Captain R.O. Grant, superintendent of the Wilmington Seacoast Railroad, reported finding human remains, including "part of a skull, with brains clinging to it, and other bones of a human body" mixed in with a load of coal that had been obtained from the Cumnock Coal Mine. After giving the matter some thought, he concluded that these were the remains of one of the miners killed in the explosion back before Christmas 1895.[50]

One of the far-reaching results of this disaster was that it led to coal mining safety legislation being passed by the North Carolina General Assembly in 1897. The new law, titled "An Act to Provide for the Inspection and Regulation of Mines," was quite comprehensive, covering a wide range of topics, including regular inspection of mines by the commissioner of labor. One thing it did not do was to provide ample compensation for the labor commissioner to implement and enforce provisions of the new law.

By February 1898, activity at the Cumnock Coal Mine had reached a record pace. The company brought in twenty-two miners from Pennsylvania to augment its workforce already on the site. By the beginning of the month, miners were bringing up one hundred tons of coal per day.[51]

Unfortunately, this flurry of activity was brought to a sudden halt on February 27, 1898, when a fire ravaged the site. All of the buildings on the surface were destroyed in the blaze. Fortunately, no one was killed or injured. Officials were sure that the fire was the work of an arsonist and put up a $100 reward for the capture of the culprit.[52]

This fire slowed the work at the mines only a little. The subterranean machinery was undamaged, and soon the work underground was moving along at a record pace. Henszey reported that the output at the mine by August 1, 1898, was three hundred tons per day and that they hoped to increase output to five hundred tons per day by the end of December. The coal was earmarked for sale to companies within the state of North Carolina, mainly railroads.[53]

Another fatal accident occurred at the Cumnock Mine, this one an explosion on the afternoon of May 22, 1900, that killed or injured twenty-five of the fifty-six miners working in the mine that day. Miners had noticed an abundance of firedamp building up in the mine in the days leading up to the blast, but work continued uninterrupted in the mine. J.D. Hart, who worked as "gas man" (the person who constantly inspected the mine looking for any signs of gas that was building up in the mine), noted in his official report:

> *Gas forms all the time—no difference in the quantity. There is a 20-inch seam of coal underneath the workings with a "black band" twenty inches thick between it and the workings. Gas suddenly accumulates in mine by forming in this seam of coal and breaking through the "black band" or from "creeps." Explosion was set in east workings, I believe in the extreme east. It was in air course. Sudden outbursts from the floor, I believe to be the cause of the accident, and there is no way to provide against it. I was not in the mine at the time of the explosion. Went down within an hour. Told them to throw water down. The mine was very hot. Went to the extreme east. Wooden props were burning. Others went down with me. I did not stay long. The air was all right, considering the explosion. Stoppings were blown out, changing the route of the air. There is a compressed air engine in*

> *the mine to pull coal up from dip to entry. Explosion at about two hundred feet from the engine. Most of the bodies were found at the engine. Don't think there was much fire, or that men were hurt much, but coming up from the dip met the bad air (after damp) and were suffocated.*[54]

Among the dead was mine superintendent John Connolly, an Englishman by birth who had worked in the coal mines near Pittsburgh before being recruited by Henszey to come south to oversee the mining operations at Cumnock. When rescuers found his body, they noticed that his back was broken in two places. He left behind a wife and three small children. His remains were sent north for burial in Pennsylvania.

J.F. Harrill, superintendent of the Cumnock Railroad, was one of the first company officials on the scene and helped direct the rescue efforts. Harrill even climbed into the mine hunting for survivors, but they rescued only five people, two of whom died from their wounds.

Harrill took the train from Cumnock to Raleigh to purchase coffins for the dead miners. He stopped long enough in Raleigh to describe to a correspondent of the *Charlotte Observer* the tragic events that transpired back in Chatham County:

> *The dead and injured, as rapidly as found by rescuers, who groped about with safety lamps in hand, were placed in the little cars and taken to the shaft and then lifted to the surface. The hair was burned off their heads and faces, necks and hands were so scorched that the skin fell off at a touch. Their ears were in most cases burned to a crisp, yet the clothing was not burned.*
>
> *One body has not been recovered, that of Sam McIntyre, colored, and it is said there is no hope of finding it. Some think McIntyre was entirely burned or blown to pieces or that he is covered with debris. The mine was not wrecked, and the costly machinery is uninjured. By midnight all the bodies were out save McIntyre's. Harrill says he worked all night. It was found that 19 men were killed by the explosion and four by the terrible after-damp. Harrill says this after-damp has a peculiar and most loathsome odor and that he felt as if his very system were saturated with it. There was a great deal of fire-damp in the mine; it appears that this has always been the case.*[55]

Initially, it was thought that a charge of dynamite had set off a firedamp explosion, but later investigation revealed that this was not the case. The most plausible explanation that mine officials and inspectors from the labor commissioner's office came up with was that the explosion was caused when firedamp was ignited by the flame of a safety lantern carried by Simeon McIntyre, whose body was the last to be recovered from the mine. Those who were in the mine at the time of the blast knew that McIntyre was near the point where the explosion began and figured that his body was blown into oblivion. But his body was found three days after the blast, along with his safety lamp.

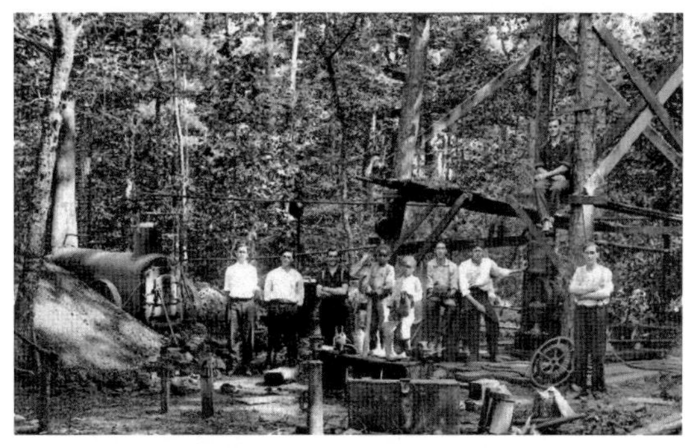

Cumnock Mine, circa 1925. *Courtesy North Carolina Archives and History.*

Investigators noticed something strange about McIntyre's lamp: "A hole was in the thick glass of his lamp, and the theory is that an excess of gas gathered suddenly from some unknown reservoir and caused the lamp to burn a hole through the glass. It is asserted that the hole is plainly broken from the interior outward."[56]

A coroner's jury made up of T.W. Segroves, Thomas J. Johnson, Oren Dowdy, J.R. Burns, R.R. Segroves and G.G. Lutterloh was convened to investigate the accident. Their report, dated May 24, 1900, noted, "We, the jurors summoned by the Coroner, have investigated the Cumnock Coal Mine disaster, and find that these twenty-one men came to their deaths by the explosion of gas and the after-damp, but how the fire originated, we can not say."[57]

Coal production declined markedly at Cumnock after the fatal explosion in 1900, and Henszey's mining operations at Cumnock did not long survive the disaster. Annual output dropped from 26,896 short tons in 1899 to 1,557 short tons in 1905. Work at the mine ceased altogether in 1905 after the mine flooded.[58]

There was a flurry of activity at Cumnock during the World War I period. A year before the United States' entrance into the war, there were efforts afoot to construct munitions and foundry operations at Cumnock. There was even talk of building a new iron furnace there that would supply iron for a proposed armor plant at Fayetteville. Nothing much came of these grand plans.[59]

Soon there was a renewed interest in Deep River coal, not just for the proposed munitions plant but also for the use of the railroads. A new company was formed in 1917 to mine the coal at Cumnock—the Cumnock

Coal Mining Company. This company was backed by both Standard Oil Company and the Norfolk and Southern Railroad. I.C. Millard of Norfolk, Virginia, was president of the new company and W.W. Brewer general manager. W.H. Hill served as superintendent. The new company cleared the mine of water and cleared out the debris left behind from the explosion in 1900. By August 1918, the Cumnock Coal Mine was once again active.

"This part of the job has been completed, and about 60 tons of coal a week are being taken out of the shaft, which is 480 feet deep," noted a contemporary account. "The work of removing rubbish and retimbering the mine with the 460 feet shaft is progressing rapidly, and by the last of September or the first of October it is expected that from 300 to 400 tons of coal will be taken from the shaft daily. They are getting a fine grade of semi-anthracite coal, which is fine for domestic use and for many by-products."[60]

The fortunes of the mine seemed to be headed in a positive direction after 1922, when it was acquired by the Erskine Ramsey Coal Company of Birmingham, Alabama. The president of the company, Erskine Ramsey, was considered one of the foremost coal men in the South. He had developed coal operations in Alabama and took a keen interest in his company's works in North Carolina. He dispatched C.S. Ramsey to head up the operations at Cumnock. A mining publication of the period noted, "C.S. Ramsey, superintendent of the Dora division of the Pratt Consolidated Coal Co., comprising the Nos. 10 and 12 openings and the Clipper Mine, has resigned and will move to Sanford, N.C., where he has accepted a position as vice-president and general manager of the Erskine Ramsey Coal Co. John Terry, mine foreman at No. 10 mine has also resigned and it is understood will be superintendent of the new operations of the Ramsey company."[61]

The Erskine Ramsey Coal Company invested quite a bit of time and resources into developing adequate housing to accommodate the workers at the mine. It built rows of houses near the mine that were given colorful names such as Frogtown, Redtown and Blacktown.[62] Despite its efforts to increase capacity and to improve the living conditions of the miners, tragedy still lurked about the mine. An explosion occurred four thousand feet below the surface at 3:00 p.m. on Wednesday, November 24, 1926, at the Cumnock Mine. Charley Shirley and Sylvester Murchison were killed during the blast, and J.H. Byerly was injured. Fortunately, the explosion was small and did not affect a larger area or set off another gas explosion. There were at least fifty other miners working in the Cumnock Mine that day, most in another section of the mine. The explosion was believed to have been started by a spark caused when two bare wires accidently crossed, igniting a small pocket of gas.[63]

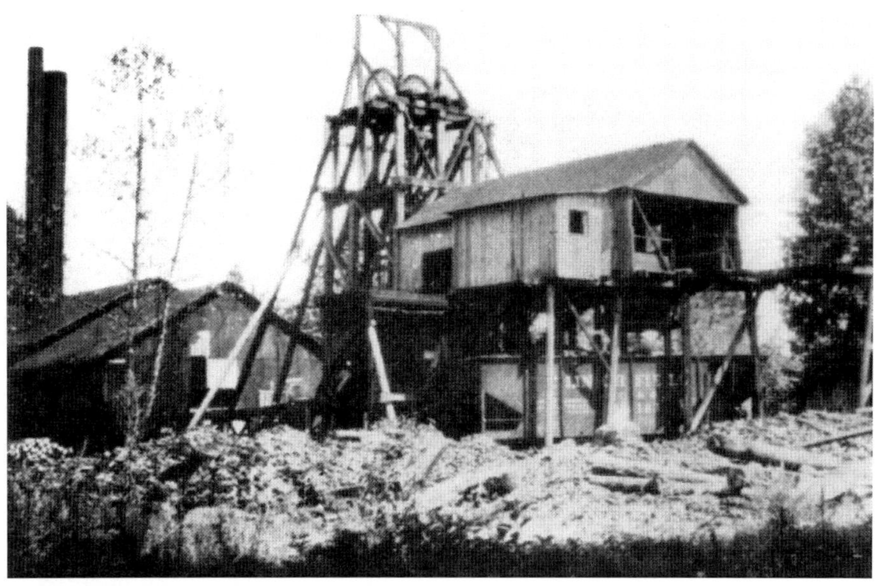

Cumnock Mine, circa 1927. *Courtesy North Carolina Geological Survey.*

Labor commissioner Frank Grist visited the mine to investigate shortly after the accident occurred. He later noted that the mine was up and running within four hours of the accident and that it would be back to running at full capacity within just a few days. "This explosion was purely local and nothing to compare with the disaster at the Carolina Mine last year," Grist stated.[64]

Grist's official investigation revealed that the explosion was caused by a short in a signal line:

> *While testing out the connections in the signal line, an arc short circuit of sparks was seen coming from the signal bell. In the vicinity of this well was known to be a small quantity of methane. The assumption is that the sparks from the signal line short circuit were of sufficient intensity to ignite the methane, causing an explosion which resulted in the death of two men. J.H. Barley* [sic], *the fire boss, was standing by the hoisting engine about 75 feet away, and was badly burned about the face. Another man, 50 or 75 feet down the shaft from the point of the explosion, was also burned about the head. Little damage was caused to the mine by the explosion, and the flames fortunately did not spread.*[65]

The mine continued operating until it was flooded by a freshet unleashed by a hurricane that struck the region in September 1928. Shortly after the

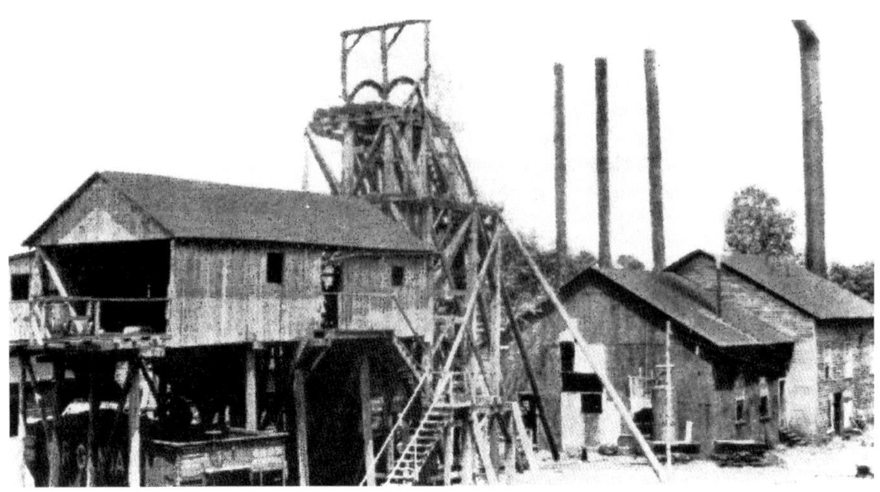

The surface works of the Cumnock Coal Mine, circa 1927. *Courtesy North Carolina Geological Survey.*

flood stopped work at the mine, the property was sold to the Carolina Coal and By-Products Company. The company already owned the Carolina Coal Mine at Coal Glen. The new company referred to its operation at Coal Glen as "Mine No. 1" and the Cumnock Mine as "Mine No. 2." Press accounts of the mine noted, "Mine number two at Cumnock was flooded by the freshet in Deep River in September but the water is being pumped out as rapidly as possible and it is thought that in a short while the shaft will be ready for mining operations to be resumed."[66]

Work of removing the water from the Cumnock Mine was nearly completed in February 1929. However, the Deep River flooded again on March 2, sending water into the mine. Miners and convicts from both the Carolina Mine and the Cumnock Mine erected barricades to hold back the floodwaters of the Deep River, but to no avail.[67]

On July 21, 1929, Ryal Berenstine, a twenty-three-year-old miner from Alabama, died from the effects of "black damp" while working in the Cumnock Coal Mine. He was working with several other miners trying to clear some flood damage in the mine. He went down to a ledge to retrieve a ruler one of his co-workers had dropped but found impossible to get because of the presence of a pocket of the deadly gas. Berenstine was sure that he could hold his breath long enough to get the ruler. However, he passed out from the effects of the gas. He was still alive but unconscious when his fellow miners got him to the surface, but they were unable to resuscitate him.[68]

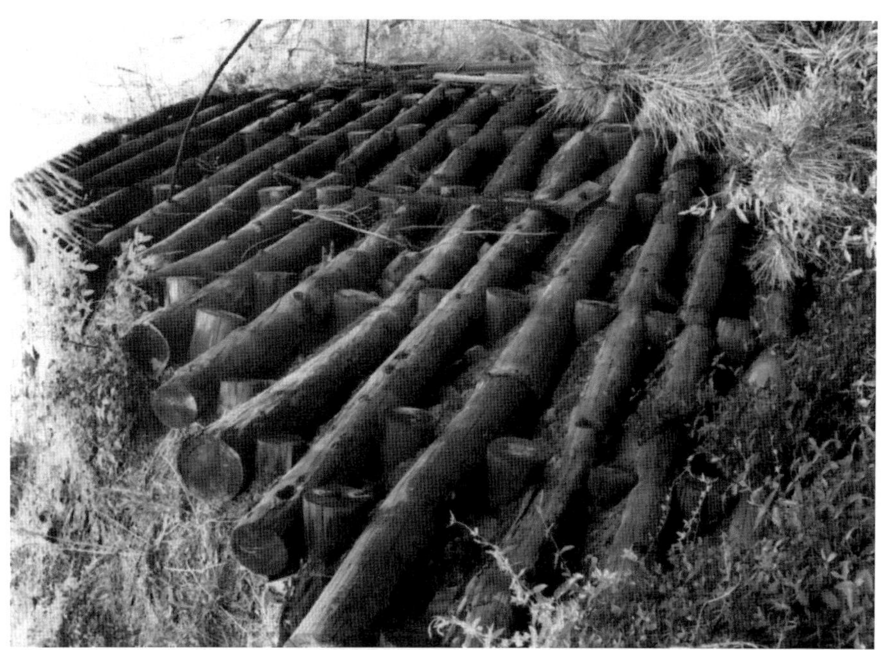

The mouth of the Egypt Coal Mine was sealed up to prevent anyone from accidently falling into the shaft. *Courtesy North Carolina Archives and History.*

Another tragic accident occurred in the Cumnock Mine on August 28, 1929, when Dan A. Moore was electrocuted. He was working in the mine when he tripped and fell, striking his head against an exposed live wire. The Carolina Coal and By-Products Company was forced by state industrial commissioner Dewey Dorsett to pay Moore's family $3,221.60. The company was also fined $3,160.00 for not having registered for coverage under the Workmen's Compensation Act, but the commissioner agreed to waive the fine if the widow and children were paid within seven days.[69]

This brought the Cumnock Coal Mine to the attention of Dan Boney, who was the state insurance commissioner. Boney sent inspectors out to examine the mine and then attempted to find a company that would insure the operations. None was found who was willing to take on the risk. The commissioner referred to the mine as a "man killer" and noted that the operations were not large enough to justify the expense needed to install the proper equipment needed to make the mine safe.[70]

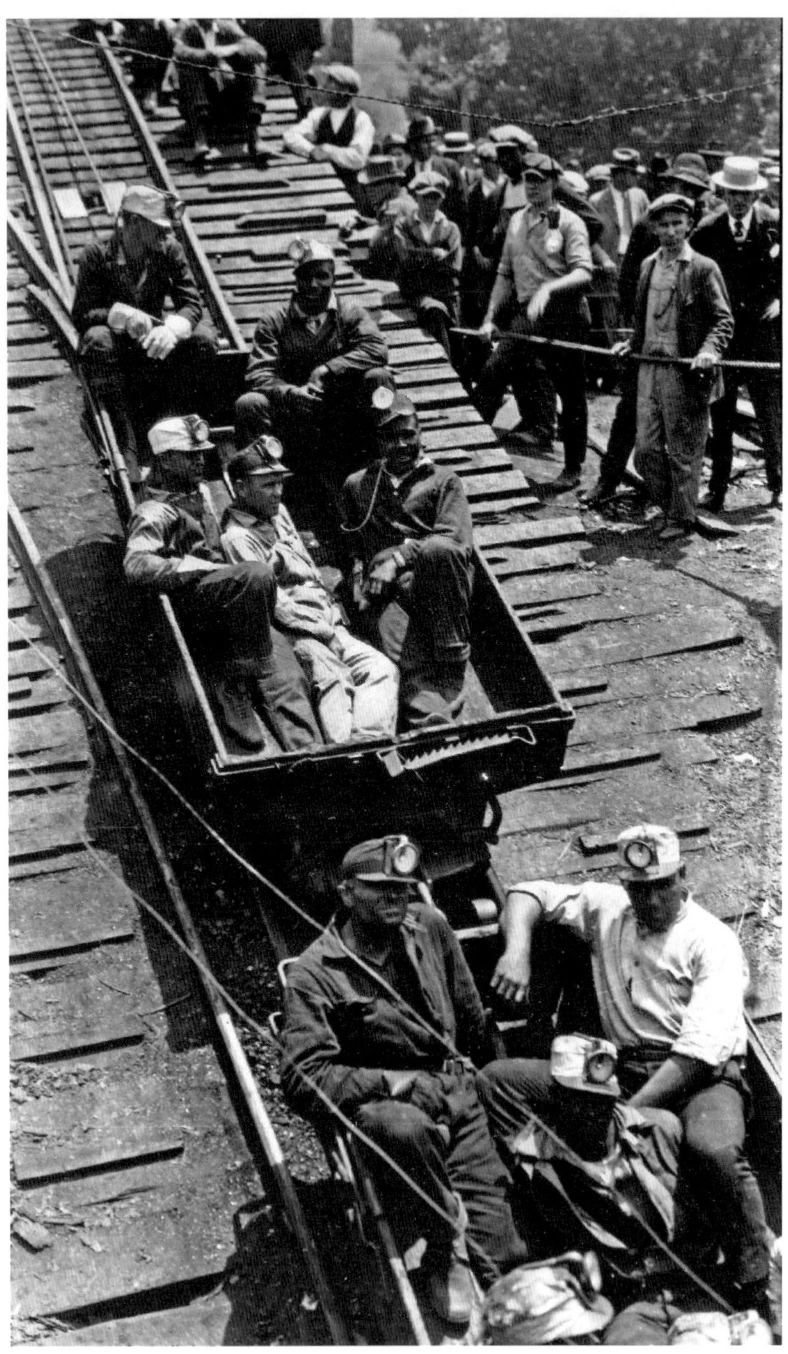

Ben Dixon McNeill captured this photo of the rescue efforts at Coal Glen in May 1925. *Ben Dixon McNeill Photographic Collection, North Carolina Collection Photographic Archives, Wilson Library, University of North Carolina–Chapel Hill.*

Chapter 5

THE DEADLY COAL GLEN MINE

The Egypt Coal Mine was not the only large-scale pit mine in the Deep River Valley with a dangerous reputation. The Carolina Coal Mine was located a mile and a half east of the Egypt Mine on the north side of the Deep River in Chatham County. Originally called Farmersville or Farmville, the mine is better known as Coal Glen, which was the name of the surrounding community, not the mine itself. Call it what you will, this is one of the most important industrial heritage landmarks in the state, as Coal Glen was the site of the deadliest industrial disaster that has ever occurred in North Carolina.

Commercial coal mining activity started at Farmville circa 1850, as miners worked surface deposits of coal found on the property. Miners did follow the coal seams underground for a short distance, but not to the extent they had on the other side of the river at Egypt. Still, the work done at Farmville prior to the war was impressive, as it is best known as being the site of the first oil refinery put in operation in North Carolina.

The coal operations drew the attention of Major T.T.S. Laidley of the North Carolina Arsenal in Fayetteville. He was dispatched to examine the timber and mineral resources in the Deep River region, and among the places he inspected was the coal operations at Farmville. He wrote of this visit, "I visited also the coal mines at Farmsville, about five miles east of Egypt. This is a surface mine, and has been worked, though it

FIRST CONSIGNMENT of North Carolina Coal.—15 tons daily expected, direct from Farmville, on Deep River. For sale by
March 21—633-tf K. M. MURCHISON.

A notice in the March 24, 1856 edition of the *Wilmington Daily Herald* discusses the first shipment of coal from the mine at Farmville via the navigational works constructed by the Cape Fear & Deep River Navigation Company. *Author's collection.*

is not at present. Permanent buildings have been erected, and a steam-engine put up for raising the coal and pumping out water from the mine; rail track and cars are provided, so that they can get to work on an extensive scale as soon as transportation is provided to take the coal to market."[71]

H.M. Chance conducted extensive exploratory work at Farmville in 1884. He was studying the various coal deposits across the state and became intrigued with the potential of the coal at Farmville. He noted that much work had been carried out prior to the war but that little had survived of the works in the fifteen years since the end of the conflict. "A considerable quantity of coal has been taken out on this Farmville property," reported Chance. "A slope was sunk on the 'main or upper bed'…before the war, and works were erected for the manufacture of illuminating oil from the bituminous slates overlying the coal. Some of the machinery may still be seen in place, but most of it was removed during the war. The hoisting machinery and boilers are still in place, but completely ruined by exposure."[72]

Chance took the opportunity of his exploration at Farmville to mine coal for the Board of Agriculture's Raleigh State Exposition. Over the four-week period, Chance and his crew utilized little more than a windlass to hoist up 165 tons of coal. Although they sank twenty-two shafts on the property for exploratory purposes, most of the coal came from two pits. Miners raised 60 tons from Pit no. 6, which was eighteen feet deep, and 105 tons from Pit no. 13, which was twenty-eight feet deep.

Chance made no mention of the previous mining fatalities in the Deep River Coal Field. However, he did include a note under "Obstacles to Successful Mining" that hinted at the dangers: "*Spontaneous Combustion (?)*.—In the Richmond coalfield great trouble has been caused by what is called spontaneous combustion. Judging from the similarity of the coals it seems possible that this same difficulty may obtain here. While this is a mere supposition, it is one that cannot safely be ignored. Mine fires are such serious disasters that every possible precaution should be taken to prevent them, and when there is any reason to suspect the possibility of spontaneous combustion, it seems essential to anticipate the danger by any precautionary measures that can be taken."[73]

There is little record of mining activity at Farmville after Chance's exploratory work in the 1880s. Whether the lack of activity was caused by legal or financial concerns is unknown. More than four decades passed from the days when Chance took the Farmville coal to the Raleigh Exposition until the return of large-scale mining operations at Farmville.

The next phase of mining activities at Farmville were well documented thanks to the presence of Bion Butler, who was one of several individuals who decided to try their luck at mining coal here in the 1920s. He explained how the finding of unusual rock intrigued a boy to the point where he conducted his own scientific experiments on it to see if anything of value could be extracted from the rock:

> *The story of the Carolina Coal mine may be interesting. It had its beginning one day when Fred M. Lane, Howard N. Butler and myself were looking over the old Taylor abandoned mine near Cumnock. Howard Butler, then a boy in his teens, found a piece of rock that excited his curiosity. He brought it home, and I told him it was oil shale. He distilled it in a crude contrivance he rigged up, and procured some oil, some gas, and ammonia water, and as I was familiar with the distillation of shales in Scotland I encouraged him to experiment further with the shale, and John N. Powell and John R. McQueen became interested with us. The four of us began some exploration, and had not gone far until we uncovered some exposed coal. We bought a lease on 1,200 acres of land, and proceeded to drill and dig and hunt for the vein. We planned to open a mine and assumed that $50,000 would permit us to mine coal. Instead of $50,000 we have now nearer three quarters of a million dollars invested, for every pound of coal taken out of the mines subsequently developed has gone back into the further development, and we have learned that it costs money to open and develop a property of this character.*[74]

In 1921, the Carolina Coal Company acquired the Farmville site and set about mining operations on a scale heretofore unseen in this part of Chatham County. The company was based in Southern Pines and was made up of J.R. McQueen, Bion H. Butler, Howard N. Butler and John Powell, with McQueen serving as president and overseeing the operations. The Carolina Coal Company worked diligently, and by 1923, miners were bringing out two hundred tons of coal per day.[75]

The company brought in the latest state-of-the-art equipment to be used in the mine. "The Carolina Coal company has received at their mines at Coal Glen on the Deep river the first coal mining machine to be brought into North Carolina. It is a Goodman shortwall, of the most modern type, electronically driven, capable of cutting down three or four carloads of coal a day. A big 11-ton hoisting engine, additional rotary, electric drive pump for underground work."[76]

Bion Butler, one of the directors of the Carolina Coal Company, was a highly regarded publicist and writer who had chronicled oil and coal operations around the world. He provided an interesting glimpse into the early work at Coal Glen:

> *The development of the Carolina company includes a slope driven on a 27-deg. pitch 800 ft. into the bed, where it begins a series of headings and workings in a seam of excellent high-volatile coal about 4 ft. thick. Assays of the coal made by various agencies show a quality comparable with the best coals of Pennsylvania—low ash, low sulphur, high heating power and from parting and impurities…*
>
> *The Carolina company's mine is planned for a daily capacity of 1,000 tons when fully developed.*[77]

On the morning of May 27, 1925, a series of explosions occurred underground in the Carolina Mine that claimed the lives of between fifty and one hundred miners and became the deadliest industrial disaster in North Carolina's history. The initial blast occurred in the fourth right entry at approximately 9:30 a.m., followed by another blast at 9:55 a.m. and a third at approximately 10:10 a.m.

Immediately after they felt the shock from the first blast, mine superintendent Howard Butler noticed smoke coming from a ventilation fan. Taking along a mechanic named Joe Richardson, the superintendent went down the slope, where they found six men lying unconscious behind the door leading into the second right. These men were carried to an area of fresher air on the mine slope.

After telephoning the men on the surface to close the explosion door, Butler and Richards continued farther into the mine. After proceeding approximately 150 feet along the slope, the second explosion occurred. Butler dropped to the floor of the slope and grabbed hold of the tracks, holding on for dear life as the pressure from the blast blew rocks and an assortment of debris by him. He survived this ordeal, although he was suffering from the effects of after-damp and several bruises to his body when he and Richardson finally reached the surface.[78]

C.M. Brown, a correspondent from a Greensboro newspaper who arrived on the scene shortly after the explosion, wrote of Butler and Richardson's foray into the mine searching for survivors: "After getting the six miners in what was thought to be a safer place, Butler began exploring a little further in the mine and then came the second explosion. Experienced miners both were and are, they threw themselves flat on the bottom of the shaft, and to this they both undoubtedly owe their lives, as the resultant rush of deadly gas would have killed them both. They crawled the distance of 1,000 feet to sunshine and safety. Richardson none the worse for his experience, although shaken, while Mr. Butler was under the care of a physician soon after his escape from the mine."[79]

Rescuers had recovered fifty-four bodies from the mine by late in the afternoon on May 30 and were working to bring out four more that had been discovered "in the inner reaches of the second right lateral and a fifth at the end of the main shaft." The fifty-ninth body was believed to be that of Joe Hudson, who died in the mine with his brother, Danny, and brother-in-law, Sam Napier.[80]

Once the rescue efforts were completed, investigators from both the federal and state governments worked with the management of the Carolina Coal Company to try to figure out what had happened. There was much talk that the explosion was the result of a firedamp explosion, much like the earlier deadly blasts at the Egypt Coal Mine. But after crawling through the mine, company officials finally came to realize that there were no signs of a firedamp explosion. Bion Butler wrote a detailed explanation of what they found:

> *The cause of the explosion was a defective shot in the fourth right entry, down near the end of the main slope. Two shots had been placed in the coal to blow it down. These shots are fired by a battery, and were both connected to the battery. One fired. The other failed. Why is not known. The powder in the shot that did not fire is to be analyzed by the government bureau of mines to see if any reason can be found for the shot failing to fire when the other one exploded. With one shot failing to*

> *fire the whole force of bringing down the coal was on the one fired, and it failed to loosen the coal. The powder blew out the tamping and flame with it. The force of the blow scattered some of the coal at the face of the hole in which the shot had been placed, and that shattered coal produced a dust under the tremendous impact, and the dust was fired by the flame from the shot, and the dust exploded. All this shown clearly by the marks of the fire on the roof of the mine, and the track of the burns on the roof and at the sides where the path of fire struck the walls, expanding as it progressed. Much has been said about gas, but so far I have seen little to indicate that gas had much to do with this explosion.*

Butler went on to explain the extensive safety equipment and ventilation systems that his company had installed in the mine in efforts to prevent an accident, especially explosions of gas:

> *Our ventilation system is one of the best we know how to provide, and it has entailed an expenditure of over $100,000. We have air at the face of every working place in the mines. We bought over $4,000 worth of lamps and equipment for charging them when the Edison electric battery lamps came out, and they are the last word in mine safety appliances. We were installing a sprinkling system to dampen the*

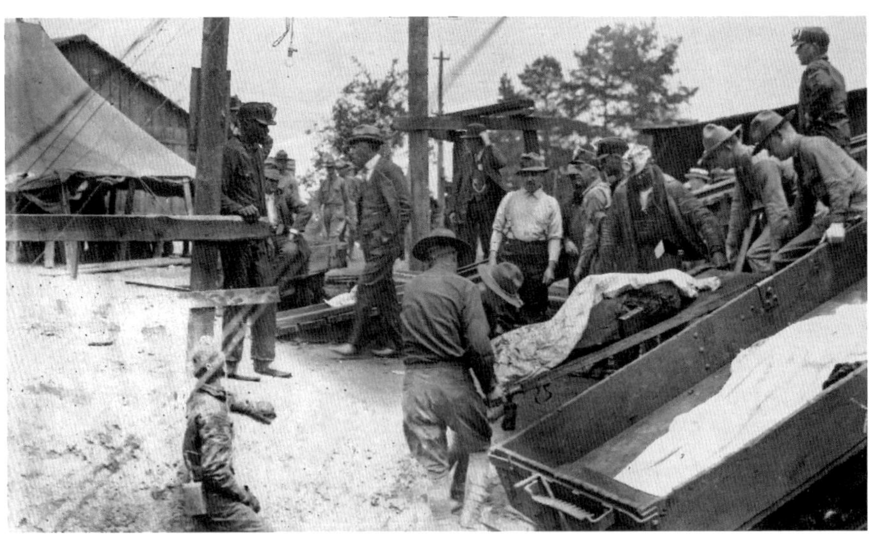

Rescuers recovering bodies from the Carolina Mine in May 1925. News and Observer *photo, courtesy North Carolina Archives and History.*

> *dust that might arise at any place in the mines, and the extent of that system no doubt kept the explosion from being more severe, although as the men were all down near the region of the first explosion where the dust from the blown out shot affected them, the water in the mines did not help them.*[81]

No one will be able to calculate the full price this tragedy had on the small community. The miners who were killed made up more than half the population of the small village of Coal Glen. The family members they left behind included thirty-eight widows and seventy-nine children.

The first steps of raising relief funds for the afflicted community came on May 30, 1925, when Governor Angus Wilton McLean issued a proclamation to the people of North Carolina urging them to come to the relief of the stricken community. The governor challenged North Carolinians to raise $35,000 by private contributions for the relief of the families of the miners killed at Coal Glen. The proclamation named J.W. Cunningham of the Banking Loan and Trust Company of Sanford as treasurer of the fund and noted that the American Red Cross would handle allocations from the fund. The funds thus raised equal $660.38 per miner and $299.15 per family member left behind.[82]

John R. McQueen took responsibility for the tragedy and went to work to personally make sure that all of the miners' families were looked after. He put the company into receivership so that all claims against the company would be on an equal footing. He then raised the money on his own personal account to pay off these claims. He tracked down survivors and performed so thorough a job that there was never a single lawsuit filed against the Carolina Coal Company.

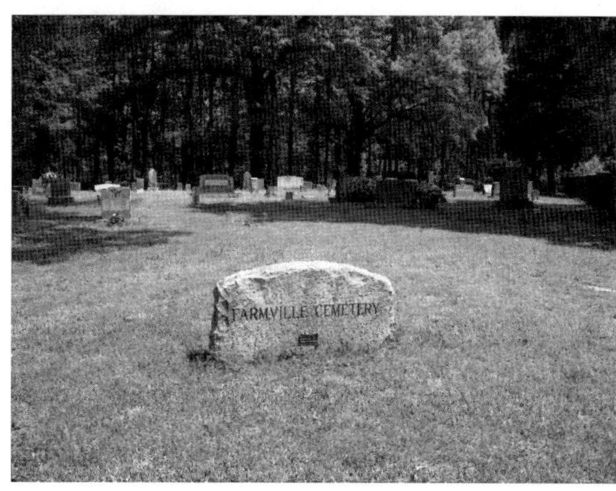

Many of the casualties from the Coal Glen disasters are buried at Farmville Cemetery. *Author's collection.*

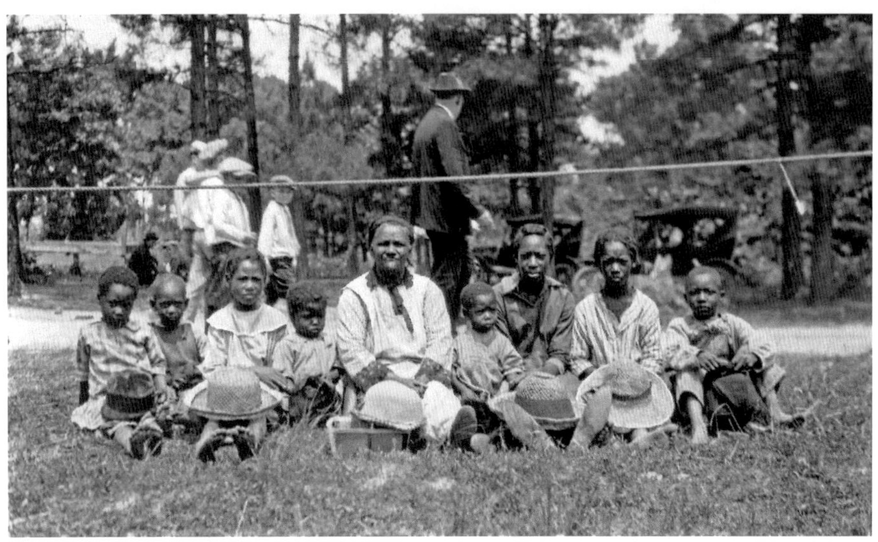

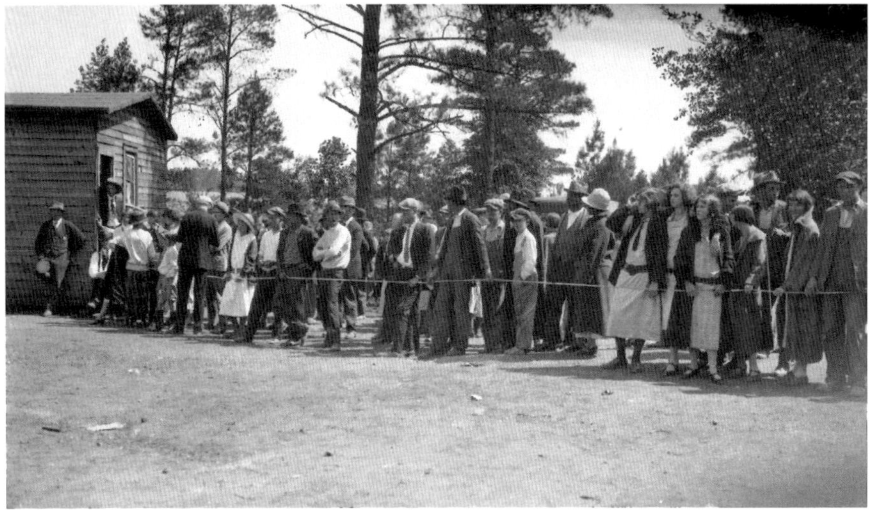

One newspaper editor wrote of McQueen's efforts, "But John McQueen is one of the most remarkable individuals in North Carolina. Where many a man would have quit, John McQueen went among his friends and borrowed money. Everybody who knows John McQueen is his friend and will do anything he asks them to do. He raised money among his friends, sought out the widows and orphans of the dead coal miners, effected a settlement in the case of every last one of them. Not a case went to court. The dependents of the coal miners were scattered over seven states. He found every last one of them."[83]

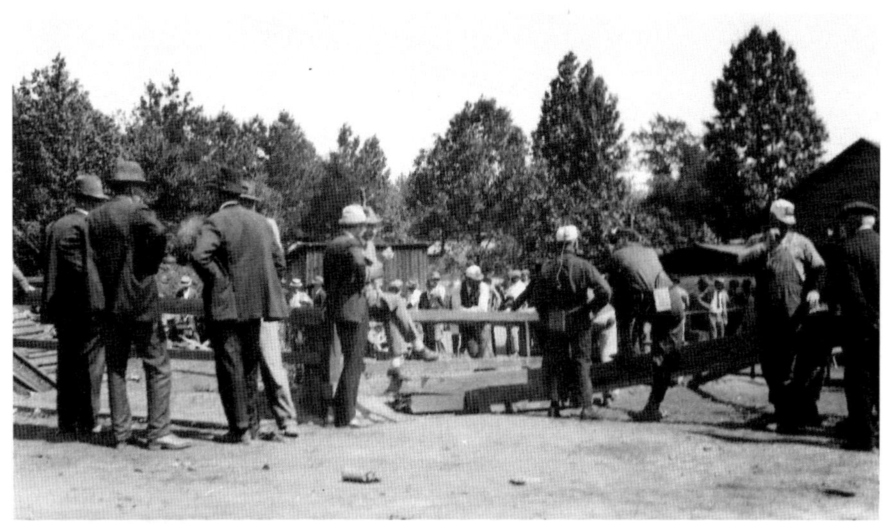

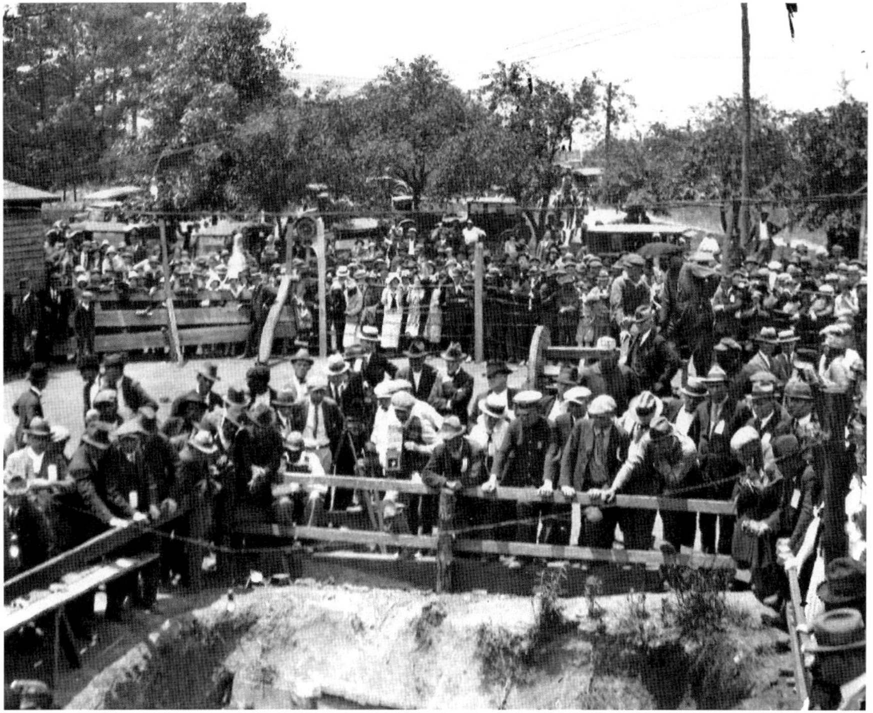

Opposite page and above: These four photographs were captured by Ben Dixon McNeill during the rescue operations at Coal Glen in May and June 1925. *Ben Dixon McNeill Photographic Collection, North Carolina Collection Photographic Archives, Wilson Library, University of North Carolina–Chapel Hill.*

ANALYSIS OF COAL.

Specimens.	FARMERSVILLE COALS.	Specific Gravity.	Volatile Matter.	Fixed Carbon.	Earthy Matter.	Ratio of fixed to volatile matter	REMARKS.
No. 1	From pit No. 1 (Sketch No. 4), lower ply, specimen recently mined.	1.416	30.91	50.77	18.32	1.64	Coke of this coal light and puffy; ashes purplish gray.
" 2	From pit No. 1, lower ply second specimen.	1.497	28.47	64.70	6.83	2.27	Coked very slowly; ashes brownish red.
" 3	From pit No. 1, upper ply.	1.547	28.06	54.78	17.16	1.95	Very intumescent; coke light; ashes purplish gray. Specific gravity is the mean of two trials.
" 4	From pit No. 2 (Farrish's old opening, see plan No. 4).	1.309	31.62	64.57	8.81	2.04	Specific gravity is the mean of two trials; ash yellowish white.
" 5	From pit No. 2 (a former analysis).	1.318	32.82	68.78	8.40	1.94	(For this analysis see "Coal Trade of British America," p. 165.)
" 6	From pit No. 5 (opened by Mr. Williams), lower ply.	1.415	30.85	63.90	5.25	2.07	Specific gravity mean of two trials: ashes nearly white, very light.
" 7	From pit No. 5, lower ply, second specimen.	1.467	30.22	47.14	22.64	1.56	Specific gravity mean of two experiments; ashes reddish, a little fused or clinkered.
" 8	From pit No. 5, lower ply, third specimen.	1.306	31.30	64.40	4.30	2.00	Specific gravity mean of two trials; coked slowly; ash nearly white.
	Averages of eight trials.	1.409	30.58	59.25	10.21	1.61	
" 9	From pits Nos. 1, 2, and 5, a mixture of 40 specimens from different piles in all those pits, pulverized together.	1.409	31.70 29.83 31.85 —— 30.79	55.61	13.77 13.83 13.70 —— 13.60	1.80	Volatile matter and ashes tested, each three times; ashes reddish grey.
" 10	Highly bituminous fossiliferous slate, found between the two plies of coal in pit No. 5.	1.791	28.60	19.45	51.95	0.65	Contains more volatile matter than fixed carbon; fragments of residue not pulverulent.

Dr. Charles Johnson provided a detailed analysis of the coal at Farmville in 1853. *Author's collection.*

The exact number of casualties from the 1925 Coal Glen disaster will never be known. The official figure of fifty-three men was put together hastily by Superintendent McQueen and released by company officials to satisfy news-hungry reporters who needed a figure for their Sunday editions. As the weeks passed and miners made their way back into the mine, they discovered remains of more previously uncounted bodies.

By the fall, company officials had revised the death toll up to seventy-one, which was the number McQueen had originally suggested based on the number of people known to be working in the mine that morning. Even this figure might be low, as many supervisors worked as independent contractors and kept track of their own numbers without reporting specific details of the makeup of independent working parties to mine management.

"These bodies have been carried to the surface and buried," explained a correspondent from the *Greensboro Record* who revisited the mine to see how work was progressing five months after the blast. "There seems to be a pocket in the depth of the mine where the odor was so bad that the officials of the mining crews believed bodies to be. From the nature of the pit and for health concerns it was considered dangerous to send men in there to search for and recover the bodies. So, one day during the past summer, this pocket was sealed with cement, and no coal will ever be taken from it in the future. This, it would seem, is a most fitting tomb for these unfortunates, who seem to have no families to claim their remains had they been carried to the upper air last May."[84]

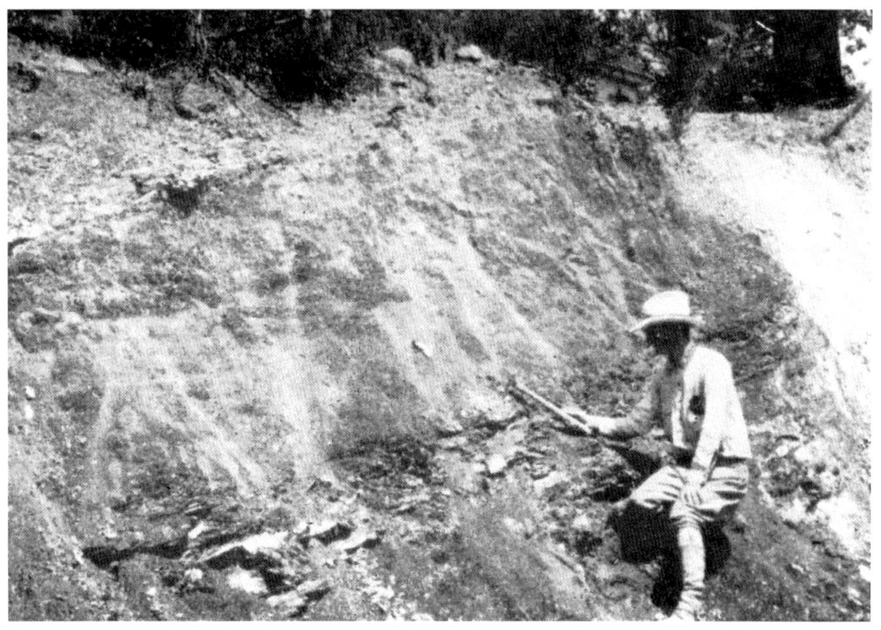

Marius Campbell examining an outcropping of the Deep River coal near Carbonton.
Courtesy North Carolina Archives and History.

Chapter 6

AFTERMATH OF THE COAL GLEN DISASTER AND THE END OF COAL MINING ACTIVITY IN NORTH CAROLINA

The Carolina Coal Mine decided to help augment its workforce with the addition of laborers from the state penitentiary in Raleigh. In April 1928, state prison superintendent George Ross Pou dispatched seventy-five prisoners to work in the coal mines. "In leasing these prisoners for mine duty Mr. Pou thinks he has found both profitable labor and safe working place for the state's wards. He says the Carolina Company has made its mining place as near perfection for safe working as any labor could be made. The superintendent has visited the mine and spent hours in it. He is sure that the state will fare well in this lease."[85]

This move was not without controversy, with many people questioning on moral grounds a program of sending miners to labor in the dangerous mines. Other states such as Tennessee utilized convicts in a similar fashion.

Pou originally sent seventy-five prisoners leased to the Carolina Coal Company. Before sending them, he personally inspected the mine to make sure it was safe for the prisoners. He was of the opinion that proper measures were in place to avoid a repeat of the 1925 disaster.[86]

On December 26, 1928, two prisoners were killed and nine injured in the Carolina Coal Mine. They were riding in the cage headed from within the mine up to the surface when the cable broke. This was undoubtedly a terrifying ride backward into the darkness of the pit as the cage plummeted over the edge and into the shaft.[87]

The death toll from this accident rose to four when two of the severely injured men died of their wounds. Those who lost their lives in this accident were James Ray of Wake County, Waddell Dorch of Durham County, Henry Simmons of Bertie County and Albert Spence of Johnston County. The wounded men who survived were John Henry Adams, Ostell Reddick, Matthew White, James Parker, George Berry, M. McEachin and Walter Bailey. These men were part of a 180-man detachment.[88]

State prison superintendent Pou was back at the Carolina Coal Mine two days after the accident. He interviewed several of the prisoners to find out not only what had happened but also if there were any danger of a similar accident occurring before authorizing the prisoners being returned to work. He pointed out that none would be forced to return into the mine who did not wish to do so: "I personally talked to every one of the 180 prisoners who are working at the mine, and almost all of them wanted to go back. One of the prisoners who was in the accident said, 'People are killed in automobiles every day, but people don't stop riding in their cars. I see no reason we shouldn't go back to work in the mine, just because we had an accident in there.'"[89]

Governor O. Max Gardner officially ended the practice of sending prisoners to work in the coal mines in 1929. By his calculations, the program his predecessor had begun fifteen months earlier had cost the lives of seven prisoners and wounded many others. In an official proclamation issued on June 19, he explained his reasoning: "As governor of this State I felt—and the prison authorities felt—that it was not the proper policy of this State to work prisoners at an extra-hazardous occupation. The withdrawal of these prisoners from the Carolina Coal Company mines will reduce the revenue of the State's prison $90,000 per year, but we felt that the question of safety of these prisoners, who are in our charge and who have to

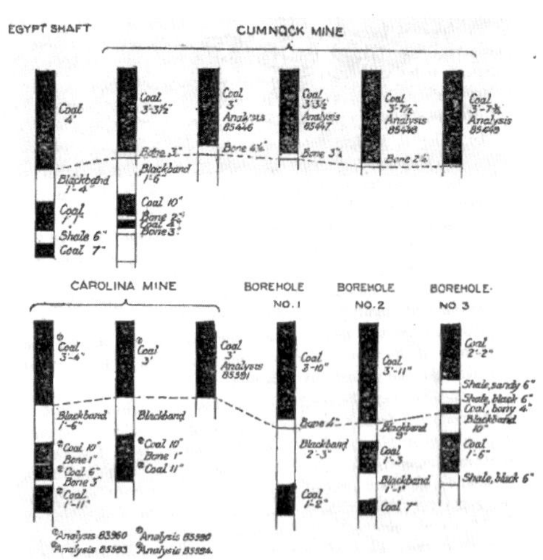

Cross-section of the coal mines at Cumnock and Coal Glen prepared by Campbell and Kimball in 1927. *Author's collection.*

work wherever they are directed to work, is of more value that the mere question of dollars and cents."[90]

Death in the mines was not restricted to the prisoners. J.W. Mullins was killed in the Carolina Mine on February 9, 1928. He was electrocuted when he came in contact with exposed wire carrying 440 volts of current. Mullins had been warning his fellow miners to be careful around the wire, which had become exposed when some rocks rolled over it and tore off some insulation during a blast. None of the miners actually saw how Mullins came in contact with the wire, but he was killed instantly.[91]

L.A. Honeycutt and his seventeen-year-old son, Elber, frequently worked together in the Carolina Coal Mine. Tragically, the lives of both father and son were taken in an accident deep beneath the surface of the earth on March 30, 1931. No one knows exactly what happened, but they were supposed to go down to the 3,600-foot level to man a water pump. A co-worker, B.A. Garner, worked at the surface, waiting for the water to start flowing, but it never did. When his shift ended, he reported that the two men had not been heard from since they entered the shaft.

Subsequent investigation revealed that the two men never made it all the way down to the 3,600-foot level. Their bodies were found 1,000 feet higher, crushed under rocks and falling debris from a cave-in. No one on duty at the mine reported any explosions that evening, so it is unclear exactly what caused the collapse that killed the father-and-son team.[92]

Tragedy continued to haunt the mine during 1931. Later that year, Fred Hanes descended 2,700 feet into the Carolina Coal Mine at 4:00 a.m. on Saturday, September 7. He was supposed to work the water pumps, but something went wrong and he was killed in an explosion. Chatham County coroner George H. Brooks gathered a jury for an official inquest but could find no cause for the deadly blast. Fortunately for the rest of the miners, the explosion was localized to the part of the mine where Hanes was working. He left behind a widow and five children.[93]

The fortunes of the Carolina Coal and By-Products Company continued to wane. Unable to raise enough capital to keep the operation going, the company was forced to file for bankruptcy. On Wednesday, February 1, 1933, Judge Thomas J. McPherson auctioned off the assets of the company to pay outstanding debts of more than $400,000, along with unpaid taxes. K.R. Hoyle, an attorney representing a group of bondholders, was the lone bidder and purchased the property for $5,000.

A newspaper article published at the time described the company's property, which included the mines at Cumnock and Coal Glen: "The holdings of the

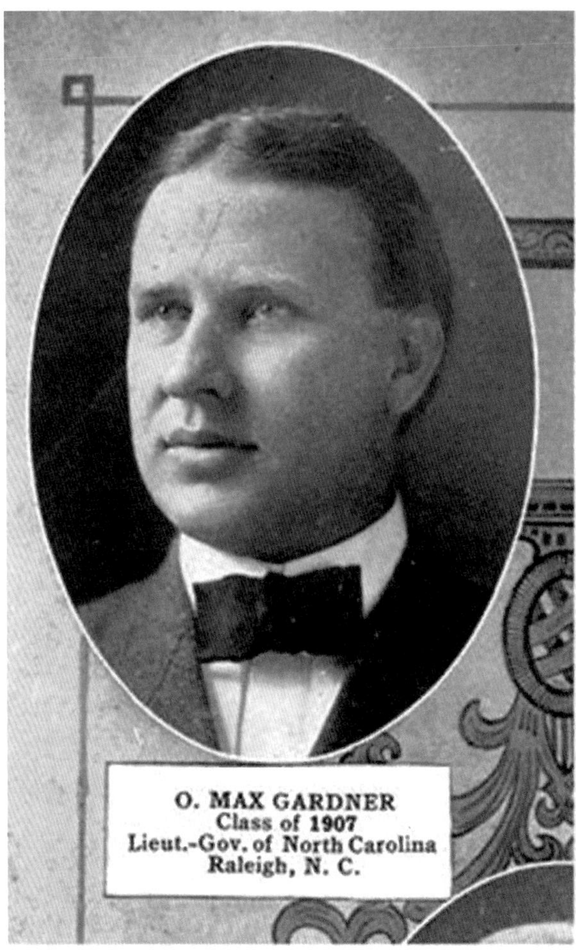

Governor O. Max Gardner was uncomfortable with sending prisoners to work in the dangerous coal mines and officially ended the practice in 1929. *Courtesy North Carolina Collection Photographic Archives, Wilson Library, University of North Carolina–Chapel Hill.*

company include more than 700 acres of coal land in fee simple in and around Cumnock and on the Chatham side of the river at Coal Glen, the mineral rights on 3,800 acres additional land in the Deep River Valley, the mines at Cumnock and Coal Glen, together with machinery and equipment, three miles of railroad track and a locomotive, and other property of one kind or another."[94]

In the fall of 1947, the Raleigh Mining Company was officially formed, its purpose to mine coal from the Deep River Coal Field. The company was a subsidiary of the Walter A. Bledsoe Coal Company of Terre Haute, Indiana. Its first order of business was to begin upgrades on the mine at Coal Glen. "Before production begins at the mine, a spokesman announced the 1,000-foot tunnel which runs 400 feet under Chatham County will be completely reworked. Rotting wooden roof supports will be replaced with steel and an armored steel liner will be added to keep the tunnel free of water."[95]

By the winter of 1949, miners from the Raleigh Mining Company were bringing up between thirty and forty tons of coal per day, with plans to increase production up to one hundred tons: "Operators more than two years ago launched a rehabilitation program at the deposit, known officially as Carolina Slope Mine in the Deep River Field. Two major steps in the rehabilitation program included installation of two coal cutting machines and placing a steel semi-circular liner through the greater part of the 1,500-foot shaft."[96]

At first, work went along well, with no accidents of note. The Raleigh Mining Company was even lauded for its safety record and awarded a special commendation by the Bituminous Casualty Corporation.[97] But as with other coal mining ventures at Coal Glen, tragedy soon struck.

Arthur Forest Devine, a coal miner from Kentucky who was working for the Raleigh Mining Company at its mine in Chatham County, died of asphyxiation brought on by a small explosion in the mine. He was still alive when his rescuers—including Robinson Cummings, James O. Bridges, Max McLeod and Frosty Glass—got to him, and it was hoped that he could be revived with a respirator that was brought in from Sanford. However, he never regained consciousness and died at the mine. He left behind a wife and four children.[98]

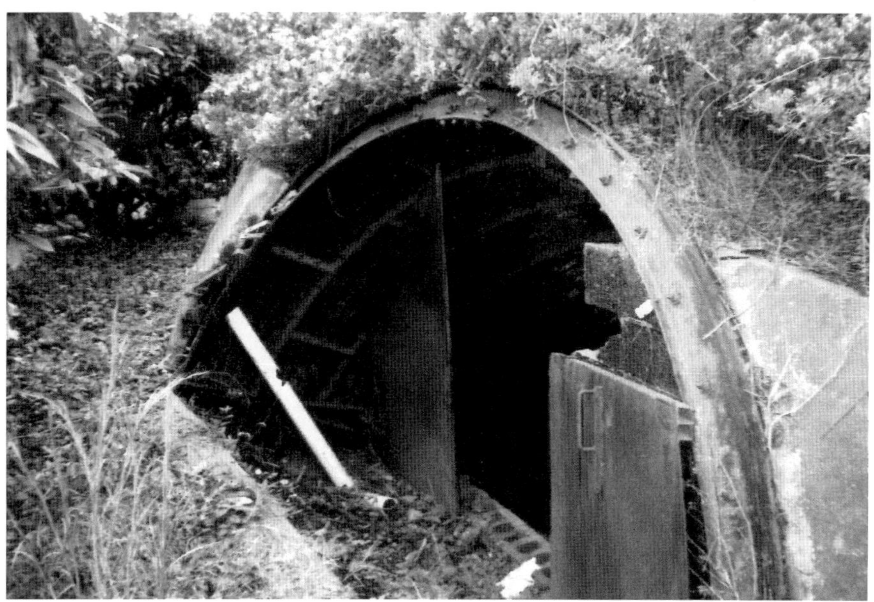

Site of the former entrance to the Coal Glen Mine as it appeared in 1997. *Author's collection.*

Next page: This map, showing the location of mineral resources of the Deep River in central North Carolina, was prepared by Captain Charles Wilkes in 1858 as part of a report to the secretary of the navy. *North Carolina Collection, Wilson Library, University of North Carolina–Chapel Hill.*

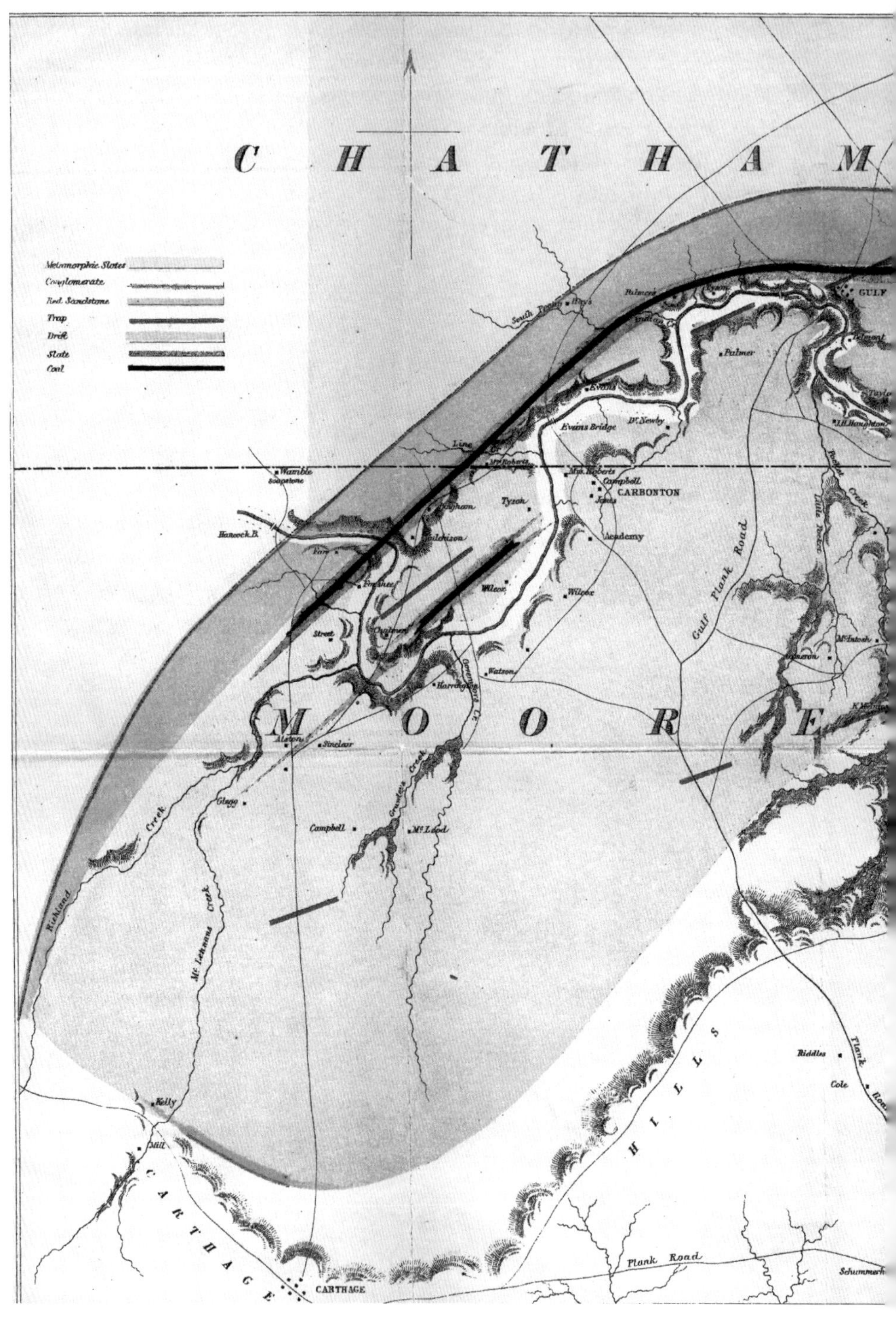
C H A T H A M
M O O R E
Metamorphic Slates
Conglomerate
Red Sandstone
Trap
Drift
Slate
Coal
GULF
Palmer
Evans Bridge
Dr. Newby
CARBONTON
Campbell
Academy
Gulf Plank Road
Wilcox
Tyson
Hancock B.
Street
Chalmers
Watson
Harrington
Alston
Sinclair
Glagg
Campbell
Mc. Leod
Richland Creek
Mc. Lennons Creek
Kelly
Mill
C A R T H A G E
CARTHAGE
H I L L S
Plank Road
Riddles
Cole
Schummerh

AFTERMATH OF THE COAL GLEN DISASTER AND THE END OF COAL MINING ACTIVITY IN NORTH CAROLINA

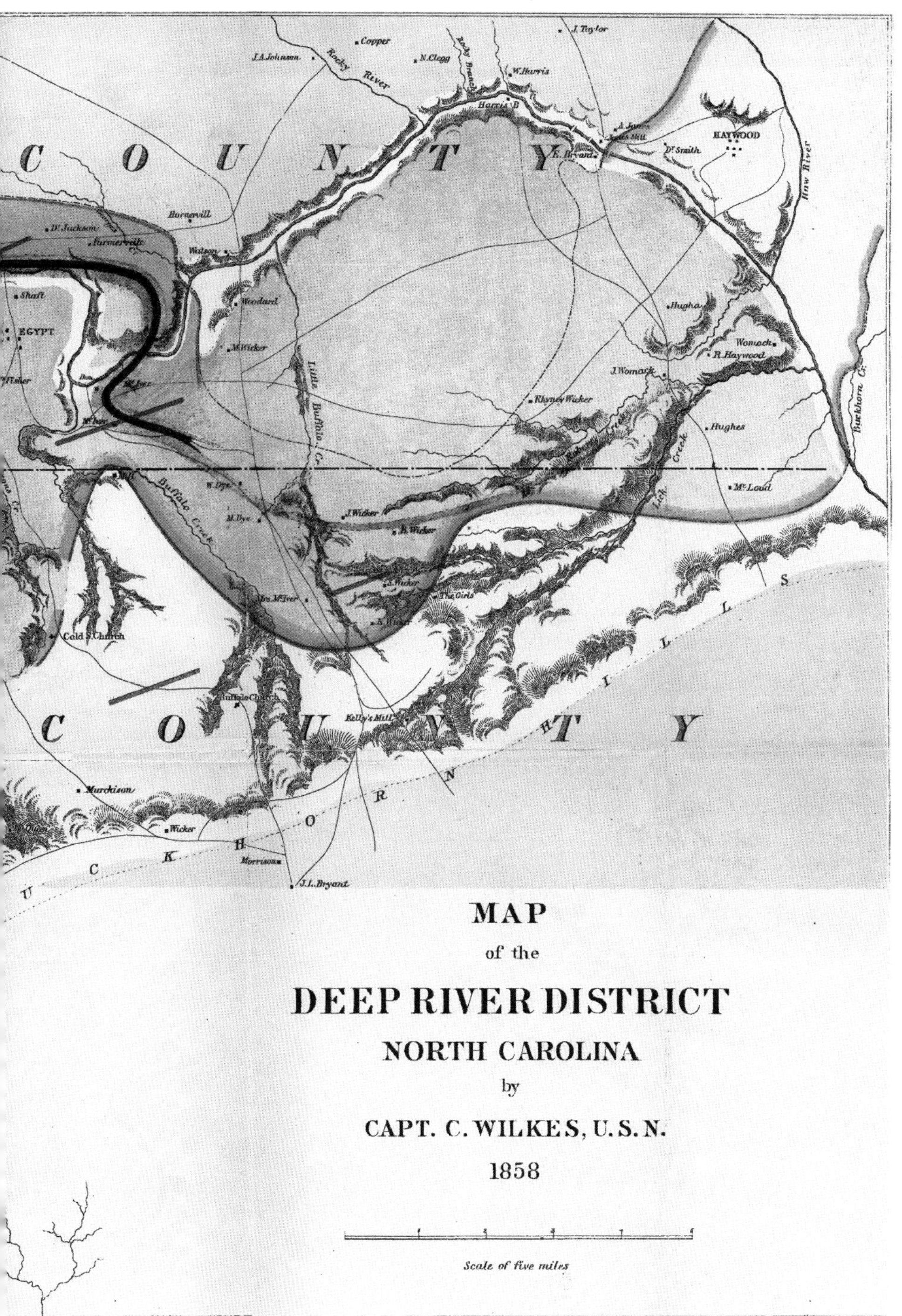

Harold Minter, a coal miner who lived near Bear Creek, was mortally wounded in the mine in March 1952. While walking in the mine, he stumbled, and to steady himself, he reached up and inadvertently grabbed a live electrical wire. He was knocked unconscious and subsequently died in the hospital.[99] Minter has the distinction of being the last person killed in a coal mining accident in North Carolina's long and sometimes tragic coal mining history.

Appendix I

A VISIT TO THE DEEP RIVER COAL FIELD IN 1856

The following story is excerpted from the North Carolina Semi-Weekly Standard, *April 26, 1856, page two. This gives a really good firsthand account of a trip to the Egypt Coal Mine, as well as other places associated with the coal operations along the Deep River.*

A Visit to the Mineral Region

One of the Editors of this paper [*Fayetteville Observer*] availed himself of an opportunity to carry into effect a purpose he has long had at heart—to visit the Coal and Iron Region of Deep River, in Chatham County. In company with a friend who had official interest in these minerals, but more especially in the timber of that region, we left here on Tuesday, and spent the following Wednesday, Thursday and Friday in riding and rambling over so much of that interesting section as was accessible within a space of some 15 or 20 miles square. And we have returned home more than ever impressed with the language of inspiration, "O Lord, how manifold are thy works! In wisdom hast thou made them all; *the earth is full of thy riches*."

The boundless wealth which there lies undeveloped, only awaits the energies of man to bring it to the light—energies which would have been long since employed if a Northern State had been favored by Providence with the rich treasures, at a point so easy of access to the markets of the world. We hope soon to see the duty performed. The work *must* be completed. The coal and iron which lie at our feet must be brought forth. The wealth and the power which such minerals bring with them to other States *must* be secured to North Carolina. And this is not to be done by any one enterprise. There is room enough for both river improvement and railroad. And so strongly do others think there is room for even more that these, a delegation of some of the most influential citizens of the State will start this day to Charleston to concert measures for another railroad to connect that city with the mines; and a part of the scheme is said to be, to continue this Charleston road to the North, by a short road from the mines to the N.C. road, a distance of 25 or 30 miles. And it is reported that Norfolk and Petersburg stand ready to do this, when Charleston shall move on the other side.

Our citizens must therefore bestir themselves, if they mean not to be outstripped in the race. The Western railroad should be pushed forward to completion with the utmost energy. If it were now completed, as it ought to have been, we are satisfied that we should have no difficulty in having the U.S. Arsenal here extended to its original design; we should have hope almost amounting to an assurance that a National Foundry would be located in this section of the State; the value of the property in this town would be increased more than the amount the road will cost; and its population and business would be correspondingly enlarged.

But let us return to our object, which was to mention what we saw on Deep River. Our first visit was to the surface mine of Messrs. Haughton, at the Gulf, which has long been known. There is no machinery here; but some hundreds of tons of coal have been taken out by hand and sold in the country around for smith's use, for which its adaptation is well known. It is delivered at the mine at 10 cents a bushel. A large heap is now lying on the ground.

We next proceeded to Egypt. This is a splendid property, even as a plantation, and is cultivated with great success under the superintendence of Wm. McClane, Esq., the Superintendent of all the interests of the Northern Company which has secured a long lease of the premises—3700 acres in all—and has, through him, erected the buildings, sunk the shaft, and put up machinery, for working this mine. The Gulf is on the Northern side of Deep River—Egypt on the Southern, about five miles distant from the Gulf. And in recrossing at Egypt, we found a free bridge in so dilapidated a condition that we thought it prudent to leave our carriage in the road and cross on foot. We beg to assure our readers in Chatham, that such a bridge does that county no credit, contrasted as it necessarily is with the beauty and order and perfection of all the works in its immediate vicinity. The county of Chatham should be excited to emulate the noble improvements of its Northern citizens.

The first objects which arrest the sight after crossing the river, are the buildings in the middle of a wide field, 1600 feet from the river, and the railroad embankment, formed of the excavated earth and stone, leading several hundred feet toward the river. On this is a temporary railroad, to be made substantial and to the river bank, so soon as the navigation shall be sufficiently forward to require it.

The Engine which drives all the machinery is a beautiful one, of 40 horse power, and seems to obey orders like a thing of life. It pumps up the water—a million gallons every 24 hours—which flows into the shaft. It conveys the workmen up and down the shaft; and brings up the excavated earth and stone and coal. The water thrown out is conveyed by pipes several hundred feet, until it reaches a steam saw mill, and there it is turned loose to carry off the saw dust which would otherwise accumulate. So that everything is turned to account. There is no waste of power.

We had no thought of going down into those lower regions of Egyptian darkness where the coal is deposited, from which we could distinctly hear the sound of the workmen engaged in blasting. But the evident safety of the descent, and the confidence inspired by the presence of Mr. McClane and his assistant, Mr. Dunn, soon induced the determination to go down. For this purpose, we speedily encased in a picturesque costume of yellow oil cloth, with a cap to correspond, and in front of the cap a small lamp was stuck—the only way in which lights are carried down or up. Placing our feet on the rim of a large bucket, hands firmly clenched in the links of a stout chain, with Mr. Dunn similarly arranged on the opposite side of the bucket and chain, the word of command was given to the engineers, and down we went. Conversing with Mr. Dunn by the way, we reached bottom in a minute or two. The bucket was immediately drawn up for our companion and Mr. McClane. And in a few minutes we were all together, admiring the wonderful works of Nature and the skill of man which made them subserve his purposes.

We were 432 feet below the surface of the ground. Below us, about fifteen feet, were workmen blasting rock for a pit which is to be 25 feet, intended to receive the drippings of water from the rock for 200 feet above. In these 200 feet there is no stream of water, only a constant dripping, which would interfere with the operations of the miners, and so it is to be collected in this lower pit (to be completed in about a week) and thence pumped 200 feet up to the "Lodgement," or reservoir, located 225 feet from the surface, into which flow the streams of water from various parts of these 225 feet, to the amount of a million gallons a day. This Lodgement extends a considerable distance on the sides of the shaft, its roof supported of course by heavy timbers. It is of itself an admirable work.

As we walked about at the bottom of the shaft, the three beautiful seams of Coal presented themselves, the upper 4 feet thick, the second 22 inches, and the lower 7 inches. Between the upper, there is slate 16 inches, and between the lower, 6 inches. We struck a pick axe into the upper seam, and down came a quantity of coal, a portion of which we brought off as a specimen of our labors as a miner.

Mr. McClane touched his lamp to a slight recess in the coal, and we were treated to a specimen of the "fire-damp," a beautiful lambent flame, which flickered over the surface of two or three feet, and then gradually went out. This fire-damp is a terrible enemy in large mines, where there is any want of care in ventilation. It sometimes produces explosions by which hundreds of lives are lost. But it is easily avoided by proper care. It will be long before there will be any danger in this mine.

The roof of the mine is smooth hard fire flay, requiring no support, so far, though as the excavations extend, the usual supports will of course be provided. The seams of coal have a slight dip, and the mine is worked upward. The advantages of this deep shaft, over the surface mining, are, the freedom from obstruction from water, and the facility of handling the coal which falls, instead of having to lift it from below. A miner with his attending laborer will get nine tons of coal a day.

After spending half an hour in the mine, we ascended in the same manner as we had gone down, except that we came up very slowly, stopping several times by the way, to examine the work. The eye had become accustomed to the darkness, so that by the aid of our small lamps everything was visible in coming up, which we had not seen at all in going down. The sides of the shaft were secured, wherever the earth or stone was loose, by stout board, firmly secured. Where the stone walls were firm, no boards were necessary.

There are 45 persons regularly employed in and about the works and farm, of whom 12 work in the pit, 4 at a time, changing every 8 hours, so as to carry on the work day and night. Most of the workmen came from the north. The whole has been accomplished so far without a single accident to life or limb—a fact as creditable to the prudence of Mr. McClane as all things about the premises are to his zeal and energy and intelligence.

Only a few hundred tons of coal have been mined so far; there being no use for any large quantity now, whilst no facilities exist for getting it to market. It is used for working the engine; and 20 or 30 tons have been sent off to the North, &c. It makes the finest possible fuel for the engine.

In the course of the excavation, four several strata of iron ore were penetrated, from 10 inches to 3 feet thick, about 6 feet in all. And fire clay, a substance indispensable for the construction of furnaces for smelting iron, was penetrated to the extent of 18 feet.

After some hours spent at this interesting spot, Mr. McClane kindly joined us in a visit to the Farmersville and Taylor mines, about 5 miles and 2 miles below Egypt, on the Northern side of the river. These are both surface mines, the former owned by a Northern company, the latter by citizens of Hillsborough, Greensborough, &c. At Farmersville, the works are all ready, buildings erected, steam engine, cable, cars, and railway provided. The Company has abundant means, we understand, and only waits the completion of the river improvement or railroad to commence operations on a large scale. A good deal of coal (probably some hundreds of tons) was taken out when their works were completed; a portion of it has been sold and some still remains on the ground. The works look rather rusty for want of use; but they will be bright enough when the way to market shall be opened.

At the Taylor mine, considerable coal has been taken out, but there are no buildings nor machinery provided as yet.

The Horneville mine, owned by Messrs. Stedman and Horne of this place [Fayetteville], and others, is still lower down the river, on the Northern side. We had not time to visit it, though we learn that a quantity of coal has been taken out, without machinery. It is the lowest mine on the North side. From thence the coal formation crosses the river and is known to Dye's, 8 miles further South, and within 35 miles of Fayetteville. How much nearer in this direction it extends has not been ascertained.

Were we must stop for to-day, for we have not time to write out the remainder of our recollections before our paper must go to press. The iron and timber and other coal regions will be the subject of a brief description in our next.

Appendix II

MEN KILLED AT COAL GLEN, MAY 27, 1925

This is the official list of the names of the miners killed during the tragic explosion at Coal Glen. The list was from a report compiled by the commissioner of labor and printing. The commissioner was on the site of the accident shortly after it occurred and interviewed several people as he tried to track down what had caused the fatal blast. This does not account for the miners whose bodies were found later once miners started working again in the mine during the summer and fall of 1925, which pushed the death toll up to seventy-one. To this day, this is the most fatal industrial disaster to ever occur in North Carolina.

Name	Age	Race	Marital Status	# Children	Occupation
Henry Alston	32	C	M	4	Miner
John Alston	24	C	S	0	Miner
F.S. Anderson	47	W	M	5	Trackman
George Anderson	60	W	M	5	Fire-boss
David Barr	19	C	M	1	Miner
Lee Buchanan	20	W	M	0	Miner
John Burgess	30	C	M	1	Mine
W.E. Byerly	39	W	M	0	Timberman
Reubin Chambliss	28	W	M	2	Miner

APPENDIX II

Name	Age	Race	Marital Status	# Children	Occupation
Wilson Chesney	59	C	M	4	Miner
June Cotton	19	C	S	0	Miner
Tom Cotton	28	W	M	1	Hoisting Engineer
J.B. Curd	40	W	M	5	Miner
C.B. Davis	35	W	M	4	Mine Foreman
N.E. Dillingham	30	W	M	0	Miner
W.D. Dillingham	27	W	M	0	Miner
H.C. Hall	34	W	M	0	Cutting Machine Oper.
Elmer Hayes	26	W	M	1	Timberman
Isaac Hayes	40	C	M	0	Miner
Lige Hill	50	C	M	5	Miner
Lee Hodges	44	C	M	0	Miner
A.L. Holland	34	W	M	4	Hoist Operator
Albert Holly	24	C	S	0	Miner
Wesley Howard	25	C	M	3	Miner
Dan Hudson	16	W	M	0	Track Helper
Joe Hudson	27	W	M	0	Miner
Will Irick	30	C	M	8	Miner
C.V. Johnson	47	W	M	3	Trackman
N.R. Johnson	25	W	M	1	Miner
Manly Lambert	25	C	M	2	Miner
J.E. Laubscher	22	W	S	0	Hoisting Engineer
A.F. Martin	38	W	M	1	Trackman
Will Moore	46	C	M	1	Miner
Jim Nabors	24	W	S	0	Miner
Sam Napier	22	W	S	0	Machine Helper
Arthur Poe	26	C	M	3	Chainer
Hollis Richardson	19	W	S	0	Chainer
Zeph Riner	22	W	M	0	Chainer
John Shaw	—	W	M	0	Miner
Charles Watson	37	C	M	1	Miner
James Williams	22	C	S	?	Miner
Robert Williams	50	C	M	2	Miner
D.J. Wilson	31	C	?	?	Timberman

APPENDIX II

Name	Age	Race	Marital Status	# Children	Occupation
Wade Wilson	33	C	?	?	Miner
C.L. Wood	25	W	?	?	Timber-helper
James Wright	25	C	S	0	Miner
P.D. Wright	49	C	S	0	Miner
Russell Wright	24	C	S	0	Miner
Tom Wright	23	C	M	6	Miner

Appendix III

NORTH CAROLINA GENERAL ASSEMBLY MINING LAW OF 1897

One of the direct results of the mining disaster at Cumnock in 1895 was the passage of a comprehensive mining law by the North Carolina General Assembly in March 1897. Although it had a few shortcomings, the law was fairly comprehensive and covered various aspects of mine safety. The following is the full text of the law, taken from the document Public Laws and Resolutions of the State of North Carolina Passed by the General Assembly at Its Session of 1897.

CHAPTER 351.
"AN ACT TO PROVIDE FOR THE INSPECTION AND REGULATION OF MINES."
The General Assembly of North Carolina do enact:

SECTION 1. That chapter 113 of the laws of 1897 is amended by adding to the duties of the commissioner of labor statistics that of "mine inspector" as herein provided for, which officer is called in this act "inspector."

SEC. 2. It shall be the duty of the inspector to examine all mines in the state as often as possible to see that all the provisions and requirements of this act are strictly observed and carried out; he shall particularly examine the works and machinery belonging to any mine, examine into the state and condition of the mines as to ventilation, circulation and condition of air, drainage and general security; he shall make a record of all examinations of mines, showing the date when made, the condition in which the mines are found, the extent record to show, to which the laws relating to mines and mining are observed or violated, the progress made in the improvements and security of life and health sought to be secured by the provisions of this act, number of accidents, injuries received or deaths in or about the mines, the number of mines in the state, the number of persons employed in or about each mine, together with all such other facts and information of public interest, concerning the condition of mines, development and progress of mining in the state as he may think useful and proper, which record shall be filed in the office of the inspector thereof as may be of public interest to be included in his annual report. In case of any controversy or disagreement between the inspector and the owner or operator of any mine, or the persons working therein, or in case of conditions of emergencies requiring counsel, the inspector may call on the governor for such assistance and counsel as maybe necessary; should the inspector find any of the provisions of this act violated, or not complied with by any owner, lessee or agent in charge, of such neglect or violation, and unless the same is within a reasonable time rectified, and the provisions of this act fully complied with, he shall Institute an action to compel the compliance therewith. The inspector shall exercise a sound discretion in the enforcement of this act. For the purpose of making the inspection and examinations provided for in this section, the inspector shall have the right to enter any mine at all reasonable times, by night or by day, but in such manner as shall not necessarily obstruct the working of the mine; and the owner or agent of such mine is hereby required to furnish the means necessary for such entry and inspection; the inspection and examination herein provided for shall extend to fire-clay, iron ore and other mines as well as coal mines.

Sec. 3. The inspector shall make such personal inspection of the mines as he may deem necessary, and his other duties will permit; he shall keep in his office and carefully preserve all maps, surveys and other reports and papers required by law to be filed with him, and so arrange and preserve the same as shall make them a permanent record of ready, convenient and connected reference; he shall compile and consolidate the reports and annually make report to the governor of all his proceedings, the condition and operation of the different mines of the state, and the number of mines and the number of persons employed in or about such mines; the amount of coal, iron ore, limestone, fire-clay or other mineral mined in this state; and for the purpose of enabling him to make such report, the owner, lessee or agent in charge of such mine, who is engaged in mining, and the owner, lessee or agent of any firm, company or corporation in charge of any fire-clay or iron ore mines, or any limestone or quarry, or who is engaged in mining or producing any mineral whatsoever in this state, shall, on or before the 30th day of November in every year, send to the office of the inspector upon blanks, to be furnished by him, a correct return, specifying with respect to the year ending on the preceding 1st day of October, the quantity of coal, iron ore, fire-clay, limestone or other mineral product in such mine or quarry, and the number of persons ordinarily employed in or about such mine or quarry below and above ground, distinguishing the persons and labor below ground and above ground. Every owner, lessee or agent of a mine or quarry who fails to comply with this section, and makes any return which to his knowledge is false in any particular, shall be deemed guilty of a misdemeanor; he shall enumerate all accidents, and the manner in which they occurred in or about the mines, and give all such other information as he thinks useful and proper, and make such suggestions as he deems important relative to mines and mining, and any other legislation that may be necessary on the subject for the better preservation of the life and health of those engaged in such industry.

SEC. 4. It is unlawful for the owner or agent of any coal mine, worked by shaft, to employ or permit any person to work therein, unless there are, to every seam of coal worked in such mine, at least two separate outlets, separated by natural strata of not less than one hundred feet in breadth, by which shafts or outlets distinct means of ingress and egress are always available to the persons employed in the mine; but it is not necessary for the two outlets to belong to the same mine, if the persons employed therein have safe, ready and available means of ingress or egress by not less than two openings. This section shall not apply to opening a new mine, while being worked for the purpose of making communications between said two outlets, so long as not more than twenty persons are employed at one time in such mine; neither shall it apply to any mine or part of a mine in which the second outlet has been rendered unavailable by reason of the final robbing of pillars previous to abandonment, as long as not more than twenty persons are employed therein at any one time. The cage or cages and means of egress shall at all times be available for the persons employed, when there is no second outlet. The escapement shafts shall be fitted with safe and available appliances by which the persons employed in the mine may readily escape in case an accident occurs, deranging the hoisting machinery at the main outlets, and such means or appliances of escape shall always be kept in a safe condition; and in no case shall an air shaft with a ventilating furnace at the bottom be construed to be an escapement shaft, within the meaning of this section. To all other coal mines, whether slopes or drifts, two such openings or outlets must be provided, within twelve months after shipments of coal have commenced from such mine; and in case such outlets are not provided as herein stipulated, it shall not be lawful for the agent or owner of such slope or drift to permit more than ten persons to work therein at any one time.

Sec. 5. The owner or agent of any coal mine, whether shaft, slope or drift, shall provide and maintain for every such mine an amount of ventilation of not less than one hundred cubic feet per minute per person employed in such mine, which shall be circulated and distributed throughout the mine in such a manner as to dilute, render harmless and expel the poisonous and noxious gases from each and every working place in the mine, and no working place shall be driven more than sixty feet in advance of a break through or airway, and all break throughs or airways, except those last made near the working places of the mine, shall be closed up by brattice trapdoors, or otherwise so that the currents of air in circulation in the mine may spread to the interior of the mine when the persons employed in such mine are at work, and all mines governed by the statute shall be provided with artificial means of producing ventilation such as forcing or suction fans, exhaust steam furnaces, or other contrivances of such capacity and power as to produce and maintain an abundant supply of air, and all mines generating fire damp shall be kept free from standing gas, and every working place shall be examined every morning with a safety lamp by a competent person or persons before any of the workmen are allowed to enter the mine. All underground entrances to any place not in actual course of working or extension shall be properly fenced across the whole width of such entrance so as to prevent persons from inadvertently entering the same. No owner or agent of any coal mine operated by a shaft or slope shall place in charge of any engine used for lowering into or hoisting out of mines persons employed therein any but experienced, competent and sober engineers, and no engineer in charge of such engine shall allow any person except such as may be deputed for such purposes by the owner or agent to interfere with it or any part of the machinery, and no person shall interfere or in any way intimidate the engineer in the discharge of his duties, and in no case shall more than two men ride on any cage or car at one time, and no person shall ride upon a loaded cage or car in any shaft or slope.

Sec. 6. All safety lamps used in examining mines, or for working therein, shall be the property of the operator of the mine, and a competent person shall be appointed for the purpose, who shall examine every safety lamp before it is taken into the workings for use, and ascertain it to be clean, safe and securely locked, and safety lamps shall not be used until they have been so examined and found safe and clean and securely locked unless permission be first given by the mine foreman, to have the lamps used unlocked. No one, except the duly authorized person, shall have in his possession a key, or any other contrivance, for the purpose of unlocking any safety lamp in any mine where locked lamps are used. No matches or any other apparatus for striking lights shall be taken into any mines, or parts thereof, except under the direction of the mine foreman. All persons violating the provisions of this section shall be deemed guilty of a misdemeanor. The mine foreman shall measure the ventilation at least once a week, at the inlet and outlet, and also at or near the face of all the entries, and the measurement of air so made shall be noted on blanks furnished by the inspector; and on the first day of each month the mine boss of each mine shall sign one of such blanks, properly filled with the said actual measurement, and present the same to the inspector, and any mining boss making false returns of such air measurement shall be deemed guilty of a misdemeanor. Every person having charge of any mine, whenever loss of life occurs by accident connected with the workings of such mines, or by explosion, shall give notice thereof forthwith, by mail or otherwise, to the inspector, and to the coroner of the county in which such mine is situated, and the coroner shall hold inquest upon the body of the person or persons whose death has been caused, and inquire carefully into the cause thereof, and shall return a copy of the finding and all the testimony to the inspector. The owner, agent or manager of every mine shall, within twenty-four hours next after any accident or explosion, whereby loss of life or personal injury may have been occasioned, send notice, in writing, to the inspector, and shall specify in such notice the character and cause of the accident, and the name or names of the persons killed and injured, with the extent and nature of the injuries sustained; when any personal injury, of which notice is required to be sent under this section, results in the death of the person injured, notice in writing shall be sent to the inspector within twenty-four hours after such death comes to the knowledge of the owner, agent or manager; and when loss of life occurs in any mine by explosion, or accident, the owner, agent or manager of such mine shall notify the inspector forthwith of the, fact, and it shall be the duty of the inspector to go himself, or send a representative at once to the mine in which said death occurred, and inquire into the cause of the same, and to make a written report, fully setting forth the condition of the part of the mine where such death occurred, and the cause which led to the same; which report shall be filed by the inspector in his office as a matter of record, and for future reference. For any injury to person or property, occasioned by any willful or intentional violation of this act, or any willful failure to comply with its provisions by any owner, agent or manager of the mine, a right of action shall accrue to the party injured for any direct damage he may have sustained thereby; and, in any case of loss of life by reason of such willful neglect of failure aforesaid, a right of action shall accrue to the personal representative of the deceased, as in other actions for wrongful death for like recovery of damages for the injury sustained.

Sec. 7. The owner, agent or manager of any mine shall also give notice to the inspector in any or all of the following cases: 1. When any working is commenced for the purpose of opening a new shaft, slope or mine, to which this act applies. 2. When any mine is abandoned, or the working thereof discontinued. 3. When the working of any mines is recommenced after an abandonment or discontinuance for a period exceeding three months. 4. When a squeeze or crush, or any other cause or change, may seem to affect the safety of persons employed in the mine, or when fire occurs. No boy under twelve years of age shall be allowed to work in any mine, and in all cases of minors applying for work, the agent of such mine shall see that the provisions of this section are not violated; and the inspector may, when doubt exists as to the age of any minors found working in any mine, qualify and examine the said minor, or his parents as to his age. In case any coal miner does not, in appliances for the safety of the persons working therein, conform to the provisions of this act, or the owner or agent disregard the requirements of this act, any court of competent jurisdiction may, on application of the inspector, by civil action in the name of the state, enjoin or restrain the owner or agent from working or operating such mines until it is made to conform to the provisions of this act; and such remedy shall be cumulative, and shall not take the place of or effect any other proceedings against such owner or agent authorized by law for the matter complained of in such action.

Sec. 8. The provisions of this act shall not apply or effect any mine in which not more than ten men are employed at the same time; but the inspector shall at all times have free ingress to such mines for the purpose of examination and inspection and shall direct and enforce any regulation in accordance with the provisions of this act that he may deem necessary for the safety of the health and lives of the miners employed therein; whosoever knowingly violates any of the provisions of this act or does anything whereby the life or health of the persons or the security of any mine and machinery are endangered, or any miner or other person employed in any mine governed by the statutes, who intentionally or willfully neglects or refuses to securely prop the roof of any working place under his control, or neglects or refuses to obey any orders given by the superintendent of a mine in relation to the security of a mine in the part thereof where he is at work and for fifteen feet back of his working place, or any miner, workman or other person who shall knowingly injure any water gauge, barometer, air course or brattice, or shall obstruct or throw open any air ways, or shall handle or disturb any part of the machinery of the hoisting engine or signaling apparatus or wire connected therewith, or air pipes or fittings, or open a door of the mine and not have the same closed again whereby danger is produced either to the mine or those that work therein, or who shall enter any part of the mine against caution, or who shall disobey any order given in pursuance of this act, or who shall do any willful act whereby the lives and health of the persons working in the mines or the security of the mine or the machinery thereof is endangered, or person having charge of a mine whenever loss of life occurs by accident connected with the machinery of such mine or by explosion, who neglects or refuses to give notice thereof forthwith by mail or otherwise to the inspector and to the coroner of the county in which such mine is situated, or any such coroner who neglects or refuses to hold an inquest upon the body of the person whose death has been thus caused, and return a copy of his findings and a copy of all the testimony to the inspector, shall be guilty of a misdemeanor, and upon conviction fined not less than fifty dollars or imprisoned in the county jail not more than thirty days or both. The owner, agent or operator of every coal mine shall keep a supply of timber constantly on hand, and shall deliver the same to the working place of the miner, and no miner shall be held responsible for accident which may occur in the mine where the provisions of this section have not been complied with by the owner, agent or operator thereof, resulting directly or indirectly the failure to deliver such timber.

SEC. 9. This act shall be in force from and after its ratification.

Ratified the 9th day of March, A.D. 1897.

Appendix IV

DEEP RIVER MINING OPERATIONS

The following article details a trip into the various mining operations along the Deep River made by a group of geology students from UNC in December 1890. The trip was organized by their professor, Dr. J.A. Holmes, who also served as state geologist. This gives a rare glimpse into the coal mining and iron operations in the area. The original article appeared in the December 2, 1890 edition of a newspaper published in Raleigh, the State Chronicle.

IRON AND COAL:
A SHORT AND INTERESTING HISTORY OF COAL AND IRON MINING
By J.A. Holmes
In the Vicinity of Egypt in Chatham County

The students in geology at the University have returned from a tour through the coal and iron regions of Chatham County under the direction of the writer.

The party, traveling in hacks and on foot, went from Chapel Hill to Gulf, thence to Egypt, Endor furnace, Farmville, Clegg copper mines, Sackville, Haywood and thence back to Chapel Hill, traversing several different geological formations, and a region of great interest and importance as connected with mining operations and river improvements undertaken at different times by State and Confederate governments and by private companies.

At Gulf, under the courteous and intelligent guidance of Mr. McIver, we examined the slopes and pits from which before, during and after the war, small quantities of coal were mined. Here also, we saw one of the several blast furnaces for the manufacture of iron created during the war with a view to supplying iron for the use of the Confederate government and home industries.

Early in the Confederacy a large increase in the facilities for the manufacture of iron was found to be necessary, and the erection of five new furnaces was begun in this region of North Carolina (Chatham County): (1) at Ore Hill (now owned by the N.C. Steel and Iron Co.; (2) at Gulf; (3) at the Tysor place 2 miles above Gulf; (4) one mile east of Egypt (Endor furnace), and (5) at Buckhorn on south side of Cape Fear river, six miles below Lockville. These furnaces were built of massive stone walls, after an old pattern, not now in use. Three of them, the Ore Hill, Endor and Buckhorn furnaces were completed and operated during the war. The Gulf furnace, perhaps the largest of them all (60 feet high and 43 feet square at the base) was never quite completed and is now falling down. The Tysor furnace is also in ruins.

The Buckhorn furnace was subsequently replaced by the more modern and much superior furnace and crushing machinery erected by Messrs. Lobdell & Co., of Delaware, some 18 years ago. This latter in turn enjoyed a short-lived activity of perhaps a couple of years and has since been lying idle. Here is now to be seen some of the heaviest machinery known in the United States.

The Endor furnace, near Egypt, is still in a fairly good state of preservation. It was located immediately on the back of Deep river so as to have the advantage of river transportation of both the ore and manufactured products. In connection with the furnace there was operated a roasting furnace, for roasting the ore, machinery for crushing the ore, small puddling furnaces and rolling mills for fashioning the iron product. After the war this property was purchased and remodeled by Messrs. Lobdell & Co., and used for a short time, but for 15 years or more it has not been used.

The Egypt coal mine was perhaps the place of greatest interest visited during the trip. But few members of the party had ever entered a coal mine, or had enjoyed the pleasure of experiencing Egyptian darkness at a depth of 500 feet underground. And as we descended the shaft, standing on the top of a huge water tank—making the descent in less than a minute, and with water pouring down on us—some thought this new experience of doubtful pleasure; and still more doubtful when, as we landed at the bottom of the shaft and started along one of the underground tunnels, the superintendent incautiously let fall the remark that there had been in past years several explosions in the mine and a number of persons killed each time. However, all was safe now, as the mine was ventilated by a large revolving fan at the top of the shaft.

The mine has had a checkered history. Some fifty years ago, coal was known to exist in the region, and the interest awakened in industrial matters as a result of its discovery exerted an important influence in the establishment, in 1850, of the State geological survey. Soon after Dr. Emmons was appointed State Geologist (1851), he examined this region and reported favorably on the problem of successful mining operations.

Dr. Mitchell, who had also examined this region, and who had long occupied a position of commanding influence in the State in matters pertaining to geology, thought the conditions unfavorable, and advised against the expenditure of money in coal mining in the region. A newspaper controversy followed between the two geologists, which is doubtless well remembered by many persons now living.

But an industrial fever pervaded this section of the State. The General Assembly and private companies were expending upward of two millions of dollars in making the Cape Fear and Deep rivers navigable between Gulf and Fayetteville. And with beds of iron ore, and coal, and fine lands, there was every reason to believe that this region was soon to be the center of great industrial activity and wealth. Dr. Emmons' favorable view as to coal prevailed. Men lost interest in farming. Lands were bought and sold for the coal and iron, or other minerals they were supposed to contain. A coal mining company was organized and under the supervision of Mr. McLean the shaft was sunk at Egypt, 1855–'57, to a depth of 460 feet. Near the bottom of the shaft was found a bed of coal 4 feet thick; just below this, a layer of slatey "black band" ½ foot thick; and below this latter, ½ foot of coal. The 4 foot bed of coal is the one being worked.

The mine was operated for several years prior to the beginning of the war, whether with financial success or not I am unable to say. During the war it was operated with varying success. The methods of mining were crude, the ventilation of the mine was poor, and times the FIRE DAMP (gas) allowed to accumulate in the mine, exploded, killing or injuring badly every man in the mine. Finally work was stopped and the mine was allowed to fill with water; and during the two decades following the timbers rotted and the machinery rusted. And could the soul of Dr. Mitchell have passed that way he would have considered the great high chimney that marked the location of the mine as a monument to one of North Carolina's disappointed hopes, to lives lost and to moneys expended against which he advised so earnestly.

APPENDIX IV

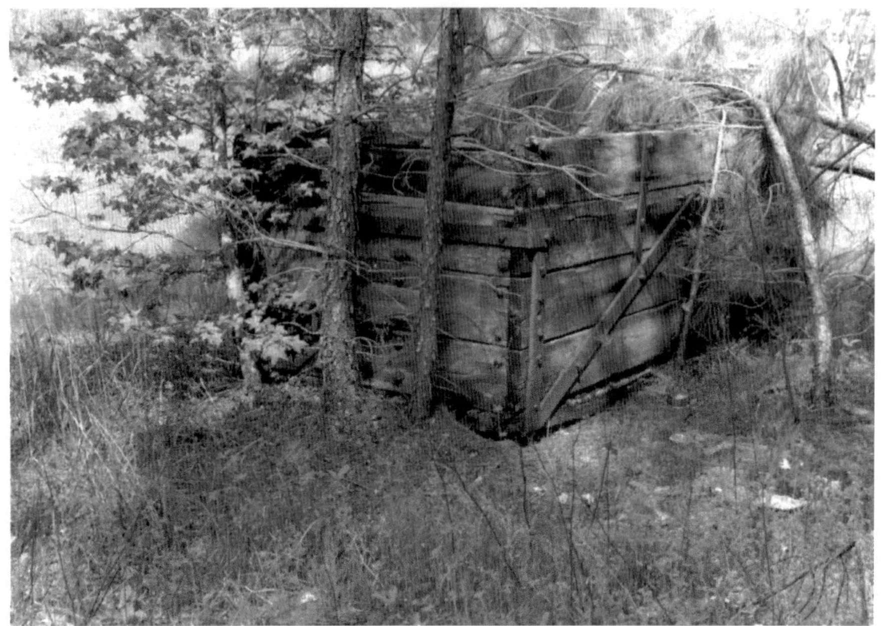

Remains of the ore crusher at the Egypt Coal Mine, circa 1969. *Courtesy North Carolina Archives and History.*

The present (Philadelphia) company, with the late Mr. Mensy as President and Mr. Gilmore as superintendent, began operations about two years ago, first clearing away the decayed timbers and drawing the water out of the mine by means of large tanks. For more than a year it has been mined on a small scale. At the present time there are about fifty men at work mining and hoisting coal, with an output of about sixty tons of coal per day, shipped to Raleigh, Durham, Greensboro, Fayetteville, Wilmington and other places in the State.

The character of the coal is usually classified as SEMI-ANTHRACITE, being neither soft coal (bituminous) nor stone coal (anthracite) but between the two. Its reputation has been injured through the ignorant and careless miners failing to separate the "black band" slate and other foreign materials from the coal. The present company is endeavoring, with encouraging results, to remedy these evils, and expect, at an early date, to be distributing over the State in much larger quantities and better quality of coal. In this it has the best wishes of the University geology class and of the people of the State at large.

GLOSSARY OF COAL MINING TERMS

This glossary is compiled mainly from the book *Glossary of Scotch Mining Terms*, published in 1886 by James Barrowman. The entries chosen have been included to give an idea of some of the nomenclature in use in the coal mining industry in the nineteenth and early twentieth centuries that would have been heard being use by the miners going about their daily tasks at the coal mines at Cumnock and Coal Glen. Since there was such a heavy influence by people from Scotland in the mines along the Deep River in central North Carolina, Barrowman's treatise on Scottish mining was deemed to be an appropriate choice.

after-damp: the mixture of gases resulting from an explosion of firedamp.
air: (noun) the ventilating current; (verb) to ventilate.
air-box: a rectangular channel made of deals for the conveyance of air for ventilation.
air-course: an underground road or passage used exclusively or chiefly for ventilation.
air-crossing: an erection or arrangement of airways whereby one air current is carried over and kept separate from another.
air-pit: a shaft used specially for ventilation.
apron: in pumping, a method of connecting parallel pump rods by filling the intervening space with short logs and fastening the whole with iron glands.
back-lye: a siding behind the shaft or other center of haulage operations.
back-mine/back-set mine: a cross-cut mine toward the dip of the strata.

backs: cleavage planes; the main joints, vertical or nearly so, by which strata are intersected.

back-splintering: working the upper of two seams backward by long wall, using the roads of the lower seam.

balance rope: a rope hung under the cage in a shaft to counterbalance the winding rope.

balls, or ball ironstone: ironstone occurring in balls or nodules.

band: a thin stratum, now used chiefly as part of the compound words *blackband*, *clayband*, *slatyband*, *roughband* and *musselband*.

bank: the surface of the ground at a pit mouth.

bare: to expose (e.g., to bare a hitch); to remove overlying earth or strata in a quarry or opencast working.

barrel: a vessel by which water is lifted by engine or windlass from sinking shafts.

barring: the wooden lining of a shaft.

bearer: a person, usually a woman or girl, who formerly carried the coal in baskets from the workings to the shaft and, in many cases, up the shaft on ladders to the surface. The bearer was usually the miner's wife or daughter.

bed: a seam or stratum.

bench: a portion of a seam that is too thick to be worked in one face.

blackband ironstone: mineral carbonate of iron, containing coaly matter sometimes sufficient in quantity for its calcination.

black coal: coal slightly burned by igneous rock.

black-damp: carbonic acid, or choke-damp.

blaes: shale; laminated clay; indurated mud.

blaes and balls: blaes with ironstone nodules imbedded.

blast: to excavate by means of gunpowder or other explosive.

blind coal: coal deprived of part of its volatile matter and that burns without smoke; anthracite coal.

blind pit: an underground shaft (i.e., a shaft that does not reach the surface); a shaft from an upper to a lower seam.

blower: the abundant and audible emission of firedamp from a fissure; the point of issue.

bossing: the holing or undercutting of a thick seam, as of limestone, the height of the undercutting being sufficient for a man to work in.

bottle-coal: gas coal.

bottom: the lowest landing in a shaft or incline; sometimes any underground landing (e.g., high-bottom, low-bottom).

brasses: iron pyrites, occurring in veins and nodules in some seams of coal.

GLOSSARY OF COAL MINING TERMS

brattice: a partition for directing the ventilating current.

break: the fracture of the strata consequent on the working out of a seam.

cabin: a shelter for workmen; an enclosed place underground used for a particular purpose (e.g., lamp cabin).

cage: the carriage or platform used to raise men and materials in a shaft.

calcining hearth: a space, usually floored with bricks, where ironstone is calcined.

cathead: inferior ironstone.

cave in: a collapse; falling in of the sides of a shaft or open working.

chain and buckets: an old method of raising water in shafts.

channel bed: a bed of gravel.

char: coke; more usually calcined ironstone.

clean coal: coal from which the dross has been separated; good or pure coal as distinguished from impure coal in a seam.

cleek: a hook. In former times, the baskets in which the coal was drawn up the shaft were attached to the rope by a cleek, and the cleek in course of time came to mean the whole organization for raising the coal from a colliery. Hence "stopping the cleek" or "stegging the cleek" (i.e., causing an interruption of the output of the coal).

coal-face: the face of the solid coal.

coal field: an extent of country having coal-bearing strata; the area of coal comprised in a winning.

collier: a coal miner.

colliery: (old form "coalery") the shaft and associated works employed in raising coal.

cribbing: a mode of lining a shaft, formerly practiced, where great pressure of water had to be withstood, by means of cribs of oak built one upon other, carefully bedded and tightly wedged.

cross-cut mine: a mine driven from one seam to another through intervening strata.

crush: breaking down of pillars by weight of superincumbent strata.

damp: gas; most frequently used in the compound words *firedamp*, *choke-damp*.

damped: Firedamp rendered inexplosive through want of air or through excess of carbonic acid gas.

damper: a sliding shutter for cutting off the draft from a furnace; a regulator.

dampscope: an instrument invented by Professor Forbes, Glasgow, for detecting firedamp.

dander: ash; clinker.

dandered coal: coal burned by, and generally mixed with, trap.

danger-board: a notice giving warning against entering a dangerous part of the workings.
darg: a day's work; the amount of mineral put out by a miner in a day.
dass: a slice or cut taken off a pillar in stooping.
daugh: soft fire-clay associated with a seam and in which the holing is usually made.
daylight mine: a mine or drift running to the surface.
deck: a platform of a cage.
derrick: a mast or jib supporting a hoisting pulley.
dialer: a mineral surveyor.
dike/dyke: a wall or vein of igneous or other rock.
dip: declivity or declination of strata.
doggar/doggart: inferior ironstone; an irregular piece of stony coal in a seam.
doubling: thickening of a seam, sometimes due to its being folded over or doubled.
drift: a mine or roadway in solid strata.
escape pit: a shaft used as a second outlet.
exploring mine: a working place driven ahead of the others to explore the field.
fire: explosive gas; firedamp.
fireclay: silicate of alumina suitable for brick making.
fire-doors: doors of the boiler furnaces.
fire-engine: name formerly given to the steam engine.
fire-hole: a space in front of boiler furnaces to hold fuel.
fire-lamp: an iron cage or grating in which a fire is kept burning in exposed places for light and heat.
fireman: a subordinate colliery official charged with the supervision of the ventilation of the workings.
fire stink, or fire styth: the burning smell of spontaneous combustion of coal.
firing-line: an appliance sometimes used in former times for clearing a room of firedamp. With a prop being set up near the face, a ring was fixed in it near the roof, and a cord or wire passed through the ring. Attaching his lamp to one end of the cord, the miner withdrew to a distance and, pulling the cord, raised the lamp to the height necessary to explode the accumulated firedamp.
forebreast: the face of a mine.
fore-hammer: a sledgehammer, commonly applied to the hammer used by a blacksmith's assistant.

forehead: the face of a mine or level.
foul: impure (e.g., foul air, foul coal).
furnace: a ventilating furnace or cube.
gas: firedamp.
gas coal: coal that yields gas of high illuminating power upon distillation, of lustreless appearance and having a conchoidal fracture.
gash: a break or opening in the strata, usually filled with sand, gravel or other loose rocks; a sand dyke.
geordie lamp: a safety lamp after the pattern of that invented by George Stephenson.
gig: a winding engine.
gig-house: a winding engine house.
gin: a machine formerly used for raising mineral in a shaft, usually driven by horses and sometimes by water power.
graith: a miner's tools; horse harness.
great coal: large pieces of selected coal; in the east of Scotland, the coal was formerly divided into four grades: great coal, chews, lime-coal and panwood.
hag: to cut as with an axe; to cut down the coal with the pick.
head-coal: formerly, the stratum of a coal seam next the roof. More usually now, the top portion of a coal seam when left unworked, either permanently or to be afterward taken down; the top coal on a loaded waggon.
heavily watered: a colliery is said to be heavily watered when the escape of water from the strata into the shaft or workings is abundant, requiring powerful pumping machinery.
heugh: a place where coal or other mineral is worked; a pit or shaft.
hewer: a miner; a pickman.
hoist: the arrangement for raising coal in bing from the level of the bing to the top of the scree.
hole: to cut in under or above a seam preparatory to dislodging it from its bed.
ill air: noxious gas, as from underground fires or choke damp; a stagnant state of the atmosphere underground.
incline: a roadway toward the rise of the strata down which mineral is brought by a self-acting arrangement with a rope or chain; an inclined roadway along which mineral is conveyed by mechanical means.
incline-bogie: a wheeled carriage for inclines, constructed so that hutches can be run on it level and be conveyed up and down.
ironstone: a stratum containing iron in combination with carbonaceous and argillaceous matter.

joints: lines of cleavage in a seam.
kettle: a cylindrical or barrel-shaped iron or wooden vessel used to raise materials in shaft sinking.
lamp: the common miner's lamp consists of a reservoir for holding oil; a "stroup" or spout for the wick; and a "hanger" or hook for carrying by or fixing to the cap.
lamp-room: a room at the pithead where safety lamps are tested, repaired, cleaned and trimmed.
lamp-station: a cabin underground where safety lamps may be opened and trimmed; a place past which no naked lights must be taken.
leading place: a working place in advance of the others, such as a heading or a level.
lime-kiln: a furnace in which limestone is calcined.
limeshells: calcined limestone.
liver rock: homogeneous sandstone devoid of planes of stratification.
lodge: a cabin at the pithead for workmen.
longwall: a system of mining by complete excavation at one working.
lum: a chimney; when the roof of a working falls to a great height in a conical form it is said to be lummed.
"men on!": a brief expression to indicate that men are in the cage to be raised or lowered in the shaft.
mine: the underground works of a colliery or metalliferous working.
mine dust: the riddlings of calcined ironstone.
mine mouth: the point where a mine leaves the surface of the ground and enters underground; an ingoing eye.
miner: a person who digs mineral other than coal. A collier digs coals, while a miner digs other minerals—the term *miner* is also used to denote anyone who digs underground.
naked light: an unprotected light; an ordinary lamp or candle.
needle: a thin copper rod kept in a shot hole while it is being tamped, and afterward withdrawn, the fuse being inserted in the hole it leaves.
outburst: a sudden accession of water or firedamp.
outcrop: the place where a seam or stratum comes to the surface of the ground, or so near it as to have no covering of hard strata.
output: the quantity of mineral raised in any period.
parrot coal: gas coal; sometimes applied to gas coal when of inferior quality.
pickling: the falling of particles from a soft roof about to collapse.
pickman: a man who digs coal with a pick; a hewer; a miner.

pillar: a block of mineral left unworked for the support of the roof; an artificial support to the roof formed of wood or stone.

pitheadman: the man in charge of the unloading of the cages and weighing of the mineral at a pithead.

pit-mouth: the surface of the ground at the top of a shaft.

plant: the machinery and fittings about a colliery.

plug and feathers: an appliance for wedging mineral, consisting of a long wedge and two strips of iron, much used for mining in rock before the introduction of gunpowder. A drill hole having been made, the plug and feathers were inserted and the plug driven by hammers between the feathers until the rock was split.

plugging: blasting by means of plug shots.

plug shot: a charge in a small hole to break up a stone of moderate size.

proud: not cohesive; coal is said to be of a proud nature when it bursts off the working face.

prove: to test the nature of a mineral field by boring or otherwise.

redd: debris; rubbish.

reek: smoke.

rhone: a wooden channel for conveying water.

roof: the stratum above a seam or working.

roof-coal: part of a seam of coal left on for a roof.

roof-stone: the stone immediately above a seam; in long-wall, the stratum of stone above the brushing.

room: a working place in stoop-and-room workings.

room and rance: a system of working with long narrow pillars; less usually a system of working with extra-large pillars and narrow rooms.

rough coal: a name sometimes given to free coal when associated with gas coal or splint coal.

safety lamp: a lamp in which the flame is isolated by wire gauze.

Scotch-gauze lamp: a safety lamp used in Scotland, the top of the lamp being wholly of wire gauze.

seam: a bed or stratum.

self-acting incline: an inclined roadway down which mineral is run by means of a rope over a pulley or drum at the top, the weight of the loaded hutches in their descent being sufficient to draw up the empty ones, which are attached to the other end of the rope.

shaft: a vertical pit.

shot: a blast of gunpowder or other explosive.

shot firer: the person appointed to fire shots in fiery workings.

skip: a basket for raising coal in a shaft; a corf.
slake: a glutinous silt adhering to the sides of deep boreholes especially in passing through fine sandstone.
soft air: a stagnant state of the ventilation.
spare-hand: a person who does odd jobs.
spur: a support to the holed coal; a portion of the holing left unholed to support the coal until the rest of the holing is completed.
squib: a straw or tube of paper filled with gunpowder for firing a shot.
steam-jet: a jet of steam discharging in an upcast shaft to produce ventilation.
stifle: noxious gas resulting from an underground fire.
stone mine: a mine driven in barren strata.
stoop and room, pillar and stall or post and stall: a system of working by which mineral is extracted from its bed in a series of galleries or rooms, leaving pillars or stoops to support the roof.
stythe: the smell of spontaneous combustion; chokedamp; after-damp.
sumping: cutting down into the floor; in sinking, cutting down at the lowest part of the shaft.
surface water: water running into underground workings from the surface of the ground.
thirl/thirling: a drift connecting two rooms; an end.
throw: a fault or dislocation in strata.
tip: the place where waggons are tilted up on a debris or other heap.
top-water: water flowing into a higher lodgment.
tow: the winding rope, which before the introduction of iron or steel ropes was made of hemp or tow.
tree: a prop.
treed: supported by props.
triping: now usually coal as it comes from the miner; formerly, and in some cases still, coal from which the large lumps have been separated. Where the latter meaning is given to the word, coal as it comes from the miner is distinguished by the term "cleek coal."
waggon: a measure of weight equal to 24 hundredweight; coal sold for delivery in carts is usually sold by the waggon of 24 hundredweight.
wandering coal: a coal seam that exists only over a small area; an irregular seam of coal.
waste, workings: in long-wall, the space from which mineral has been extracted; in stoop and room, the area formed into pillars. The area from which the pillars have been removed is called stooped waste.

water-balance: an arrangement by which a descending tank of water raises mineral in a shaft by a rope passed over a pulley. Sometimes used where water is abundant and can be run off at the pit bottom by means of a day-level.

water barrel: a barrel for drawing water in sinking pits.

water blast: the violent escape of confined air up through water in a shaft; the discharge of water down a shaft to produce or quicken ventilation.

wedging-crib: a ring of wood or iron bedded on rock in a shaft and on which tubbing rests, the escape of water under it being prevented by the space behind it being filled with pieces of wood inserted endways and tightly wedged with wooden wedges.

weight: the pressure of the upper strata on the coal face by which, if the working is systematically carried on, the excavating of the mineral is facilitated.

white-damp: carbonic oxide.

windlass: a rope roll or drum with handles, now used in sinking shallow pits and formerly used in winding from shallow pits.

winged pillars: pillars that have been reduced in size.

workings: excavations; places from which mineral has been worked; the area within which excavations have been made.

wyper-shaft: a rocking shaft usually actuated by the eccentric rod of a steam engine and giving motion to the slide valve.

NOTES

Chapter 1

1. Hall and Snelling, *Coal-Mine Accidents*, 5.
2. Humphrey, *Historical Summary of Coal-Mine Explosions*, 23–24.
3. *Wilmington Morning Star*, "Gave His Life for Cause," July 14, 1915, 1.

Chapter 2

4. Campbell and Kimball, *Deep River Coal Field*, 14.
5. Jackson, *Report of the Coal Land*, 24.
6. *Fayetteville Observer*, "A Visit to the Mineral Region," April 21, 1856, 2.
7. Andrew, *Coal Mines*, 87–88.
8. *North Carolina Semi-Weekly Standard*, "A Visit to the Mineral Region," April 26, 1856, 2.
9. *Fayetteville Observer*, "The Egypt Disaster," March 2, 1857, 2.
10. Ibid., "Dreadful Accident," March 2, 1857, 2.
11. Ibid., "The Egypt Fire Damp Explosion," April 20, 1857, 1.
12. *Charleston Mercury*, "Raleigh, NC," April 15, 1857, 2.
13. Ibid.
14. Ibid.
15. Ibid., "The Egypt Mine in Chatham," November 16, 1857, 1.

16. Ibid.
17. Hadley, *Story of the Cape Fear and Deep River.*
18. Tolbert, *Papers of John Willis Ellis*, 456.

Chapter 3

19. *Carolina Observer*, "Coal," October 28, 1861, 3.
20. A.C. Murdock, "A Visit to the Coalfields," *Hillsborough Recorder*, November 27, 1861, 3.
21. *Carolina Observer*, "Coal," December 16, 1861, 1.
22. *Fayetteville Observer*, "Egypt Coal Mine," January 26, 1863.
23. *Carolina Observer*, "The Egypt Coal Mine," December 22, 1862, 1.
24. Knapp and Glass, *Gold Mining in North Carolina*, 19–20.
25. *Fayetteville Observer*, "The Explosion at Egypt," November 16, 1863, 1.
26. *Hillsborough Recorder*, "Fire Damp Explosion at Egypt," November 27, 1863, 3.
27. *Richmond Examiner*, "City Intelligence," February 16, 1864, 1.
28. Ibid.
29. *Richmond Whig*, "Yankee Deserters to Work in Coal Mines," February 23, 1864, 3.
30. *Charleston Mercury*, "In the Coal Mines," February 27, 1864, 2.
31. *Richmond Whig*, March 4, 1864, 3.
32. Day, *Mineral Resources of the United States.*
33. *Carolina Observer*, "Coal," November 17, 1862, 3.
34. C.B. Mallett Papers, Southern Historical Collection, University of North Carolina–Chapel Hill.
35. *Supplement to the Official Records of the Union and Confederate Armies*, part 2, Record of Events, vol. 43, serial no. 55, 807–49.

Chapter 4

36. Chance, *Report Upon an Exploration.*
37. Reinemund, *Geology of the Deep River Coal Field*, 89.
38. Harrison, *Encyclopedia of Contemporaneous Biography.*
39. *Chatham News*, "The Egypt Property," November 7, 1889, 2.

40. Ibid.
41. *Chicago Herald*, "Hours of Deadly Peril," February 27, 1890, 9.
42. *Anderson Intelligencer*, "Facing Death in Darkness," March 6, 1890, 1.
43. *Charlotte Observer*, September 30, 1893, 2.
44. Hadley, Goerch and Strowd, *Chatham County*, 192.
45. McLaren, *Shell Guide to Scotland*, 155.
46. *Charlotte Democrat*, "From Our Raleigh Correspondent," July 19, 1895, 3.
47. *Charlotte Observer*, "Temple Doors," December 25, 1895, 1.
48. *Sanford Express*, "Forty-Three Dead," December 19, 1895, 1.
49. *New York Times*, "Two-Score Miners Dead," December 20, 1895, 1.
50. *Chatham Record*, "Wilmington Messenger," March 12, 1896, 3.
51. *Charlotte Observer*, "The Cumnock Mine," February 11, 1898, 2.
52. Ibid., "Secretary Alexander Quits," March 1, 1898, 8.
53. Ibid., "Henszey-Langdon Quarrel," July 30, 1898, 8.
54. *Fourteenth Annual Report of the Bureau of Labor and Printing*, "Cumnock Mine Disaster," 357–69.
55. *Charlotte Observer*, "Details of the Explosion," May 24, 1900, 8.
56. *Macon Telegraph*, "Cumnock Mine Horror," May 26, 1900, 2.
57. *Fourteenth Annual Report of the Bureau of Labor and Printing*, "Cumnock Mine Disaster," 367.
58. Day, *Mineral Resources of the United States*.
59. *Greensboro Daily News*, "The Proposed Development at Cumnock," April 6, 1916, 4; see also Alfred M. Myrover, "Opening of Coal Mines Means Much to State," *Charlotte Observer*, December 14, 1916, 14.
60. *Greensboro Daily News*, "Coal Is Being Mined at the Cumnock Mine," August 8, 1918, 8.
61. *Coal Age*, "Alabama," November 16, 1922, 320.
62. *Greensboro Daily News*, "Mine at Cumnock to Increase Output," April 17, 1927, 20.
63. *Winston-Salem Journal*, "Two Killed, One Injured at Sanford," November 25, 1926, 1.
64. *Greensboro Daily News*, "Cumnock Explosion Caused Little Delay," December 4, 1926, 19.
65. *Greensboro Record*, "Report Made on Mine Explosion," December 7, 1926, 1.
66. *Greensboro Daily News*, "New Concern Takes Over 2 Coal Mines," December 9, 1928, 29.
67. Ibid., "Cumnock Coal Mine Once More Flooded," March 13, 1929, 32.
68. Ibid., "Black Damp Kills Miner at Cumnock," July 22, 1929.

69. M.R. Dunnagan, "Must Pay Widow for Man's Death," *Winston-Salem Journal*, December 6, 1929, 2.
70. Ibid.

Chapter 5

71. T.T.S. Laidley, "Timber and Minerals of the Deep River Country, North Carolina," 34th Congress, 1st Session, House of Representatives, Executive Document No. 109, 1856.
72. Chance, *Report Upon an Exploration*, 27.
73. Ibid.
74. Bion Butler, "Backfire Cause of Coal Glen Mine Disaster," *Charlotte Observer*, June 25, 1925, 10.
75. Reinemund, *Geology of the Deep River Coal Field*, 90.
76. *Charlotte Observer*, "Has Only Coal Mining Machine in the State," November 20, 1922, 7.
77. Bion Butler, "North Carolina Again Among Coal States," *Coal Age*, November 23, 1922, 337.
78. Humphrey, *Historical Summary of Coal-Mine Explosions*, 102–3.
79. C.M. Brown, "Recover 12 Bodies from the Mine," *Greensboro Record*, May 28, 1925, 1, 11.
80. *Greensboro Record*, "May Find Other Bodies Covered by Debris," May 30, 1925, 1.
81. Butler, "Backfire Cause of Coal Glen Mine Disaster," 10.
82. Corbitt, *Public Papers and Letter*, 132–33.
83. *Sanford Express*, "Coal Mines," March 21, 1929, 1.
84. *Greensboro Record*, "Seventy-One Miners Met Death in Coal Glen Mine Last May," October 18, 1925, 8A.

Chapter 6

85. *Greensboro Daily News*, "Prisoners to Work in Cumnock Mines," April 5, 1928, 11.
86. *Greensboro Record*, "Prisoners to Work in the Cumnock Mine," April 5, 1928, 11.

87. *Greensboro Daily News*, "Two Prisoners Meet Death at Carolina Coal Mine," December 27, 1928, 14.
88. *Greensboro Record*, "4 Die, 7 Hurt in Coal Mine Near Sanford," December 27, 1928, 1.
89. Ibid., "Prisoners Go Back to Jobs in Coal Mine," December 30, 1928, 4.
90. Corbitt, *Public Papers and Letter*.
91. *Winston-Salem Journal*, "Miner Instantly Killed by Wire," February 9, 1928, 4.
92. *Richmond Times Dispatch*, "Two Bodies Found Following Cave-in at N.C. Coal Mine," March 31, 1931, 12.
93. *Greensboro Daily News*, "Killed in Mine," September 8, 1931, 3.
94. *Sanford Express*, "Cumnock Mine Is Sold for Small Amount of $5000," February 2, 1933, 1.
95. *Greensboro Daily News*, "Work of Repairing Chatham's Tunnel Gets Under Way," October 3, 1947, 11.
96. Ibid., "Deep River Mine Produces 40 Tons of Coal Per Day," February 8, 1949, 1.
97. Ibid., "Company Receives Safety Flag," June 24, 1951, 52.
98. Ibid., "Man Is Killed in Lee Mine Accident," January 24, 1952, 6.
99. Ibid., "Accident Kills Miner," March 13, 1952, 6.

SELECT BIBLIOGRAPHY

Andrew, Roy. *The Coal Mines.* Cleveland, OH: Robinson, Savage & Company, 1876.

Campbell, Marius R., and Kent W. Kimball. *The Deep River Coal Field of North Carolina.* North Carolina Geological and Economic Survey, Bulletin No. 33 (1923).

Chance, H.M. *Report Upon an Exploration of the Coalfields of North Carolina.* Raleigh, NC: P.M. Hale, 1885.

Corbitt, David Leroy, ed. *Public Papers and Letters of Oliver Max Gardner, Governor of North Carolina 1929–1933*. Raleigh, NC: Council of State, 1937.

Day, David T. *Mineral Resources of the United States Calendar Year 1905.* Washington, D.C.: Government Printing Office, 1906.

Emmons, Ebenezer. *Geological Report of the Midland Counties of North Carolina.* New York: George P. Putnam & Company, 1856.

Fourteenth Annual Report of the Bureau of Labor and Printing of the State of North Carolina for the Year 1900. "Cumnock Mine Disaster." Raleigh, NC: Edwards & Broughton and E.M. Uzzell, State Printers, 1901.

Fowler, Malcolm. *They Passed This Way*. Lillington, NC: Harnett County Centennial Committee, 1955.

Hadley, Wade. *The Story of the Cape Fear and Deep River Navigation Company, 1849–1873*. Siler City, NC: Chatham County Historical Association, 1980.

Hadley, Wade, Doris Goerch Horton and Nell Craig Strowd. *Chatham County, 1771–1971*. Durham, NC: Moore Publishing Company, 1971.

SELECT BIBLIOGRAPHY

Hall, Clarence, and Walter O. Snelling. *Coal-Mine Accidents: Their Causes and Prevention.* Washington, D.C.: Government Printing Office, 1907.

Harrison, Mitchell C. *An Encyclopedia of Contemporaneous Biography.* Vol. 1. New York: New York Tribune, 1902.

Holmes, J.A. "Iron and Coal." *State Chronicle,* December 2, 1890, 2.

Humphrey, H.B. *Historical Summary of Coal-Mine Explosions in the United States, 1810–1958.* Bureau of Mines, Bulletin 586 (1960).

Jackson, Charles T. *Report of the Coal Land of Egypt, Belmont, Evans, Palmer and Wilcox Plantations.* New York: George E. Nesbit and Company, 1853.

Johnson, Walter R. "The Coal Lands of the Deep River Company in North Carolina, with Analysis of the Minerals." *Mining Magazine* 1, no. 4 (1853): 352–65.

Knapp, Richard F., and Brent D. Glass. *Gold Mining in North Carolina.* Raleigh, NC: Division of Archives and History, 1999.

McLaren, Moray. *The Shell Guide to Scotland.* London: Ebury Press, 1967.

Reinemund, John A. *Geology of the Deep River Coal Field North Carolina.* United States Geological Survey Professional Paper No. 246. Washington, D.C.: U.S. Government Printing Office, 1955.

Tolbert, Noble J., ed. *The Papers of John Willis Ellis.* Vol. 2, *1860–1861.* Raleigh, NC: State Department of Archives and History, 1964.

Willcox, George. *A History of the House in the Horseshoe.* Wilmington, NC: Historical Research Services, 1999.

ABOUT THE AUTHOR

John Hairr is a historian and writer who has written extensively about the region on a wide range of topics, including hurricanes, sharks, rivers, exploration and marine mammals. He is perhaps best known for his critically acclaimed work, *Col. David Fanning: The Adventures of a Carolina Loyalist*, the first and only biography to deal with the controversial Loyalist leader. John's work has appeared in publications such as the *Encyclopedia of North Carolina*, *The Journal of the North Carolina Academy of Science*, *North Carolina Historical Review*, *Fortean Times*, *Mercator's World*, *South Carolina Wildlife* and *Our State*. He has received many accolades for his work, including being named Historian of the Year in 1996 by the North Carolina Society of Historians. In addition to his literary pursuits, John is, among other things, the former site manager of House in the Horseshoe State Historic Site.